T0331702

SYSTEMS BIOLOGY: SIMULATION OF DYNAMIC NETWORK STATES

Biophysical models have been used in biology for decades, but they have been limited in scope and size. In this book, Bernhard Ø. Palsson shows how network reconstructions that are based on genomic and bibliomic data, and take the form of established stoichiometric matrices, can be converted into dynamic models using metabolomic and fluxomic data. The Mass Action Stoichiometric Simulation (MASS) procedure can be used for any cellular process for which data is available and allows a scalable step-by-step approach to the practical construction of network models. Specifically, it can treat integrated processes that need explicit accounting of small molecules and proteins, which allows simulation at the molecular level. The material has been class-tested by the author at both the undergraduate and graduate level.

All computations in the text are available online in MATLAB® and MATHEMATICA® workbooks, allowing hands-on practice with the material.

Bernhard Ø. Palsson is The Galletti Professor of Bioengineering and Adjunct Professor of Medicine at the University of California, San Diego.

Systems Biology

SIMULATION OF
DYNAMIC NETWORK STATES

Bernhard Ø. Palsson
University of California, San Diego

CAMBRIDGE
UNIVERSITY PRESS

University Printing House, Cambridge CB2 8BS, United Kingdom

One Liberty Plaza, 20th Floor, New York, NY 10006, USA

477 Williamstown Road, Port Melbourne, VIC 3207, Australia

314-321, 3rd Floor, Plot 3, Splendor Forum, Jasola District Centre, New Delhi - 110025, India

79 Anson Road, #06-04/06, Singapore 079906

Cambridge University Press is part of the University of Cambridge.

It furthers the University's mission by disseminating knowledge in the pursuit of education, learning and research at the highest international levels of excellence.

www.cambridge.org
Information on this title: www.cambridge.org/9781107001596

First published 2011
Reprinted 2014

A catalogue record for this publication is available from the British Library

ISBN 978-1-107-00159-6 Hardback

Additional resources for this publication at http://systemsbiology.ucsd.edu

TO
ERNA AND PÁLL

Contents

Preface   *page* xi

1   Introduction 1
    1.1   Biological networks 1
    1.2   Why build and study models? 5
    1.3   Characterizing dynamic states 6
    1.4   Formulating dynamic network models 7
    1.5   The basic information is in a matrix format 11
    1.6   Studying dynamic models 13
    1.7   Summary 16

2   Basic concepts 17
    2.1   Properties of dynamic states 17
    2.2   Primer on rate laws 20
    2.3   More on aggregate variables 25
    2.4   Time-scale decomposition 28
    2.5   Network structure versus dynamics 31
    2.6   Physico-chemical effects 34
    2.7   Summary 36

PART I. SIMULATION OF DYNAMIC STATES

3   Dynamic simulation: the basic procedure 41
    3.1   Numerical solutions 41
    3.2   Graphically displaying the solution 43
    3.3   Post-processing the solution 49
    3.4   Demonstration of the simulation procedure 51
    3.5   Summary 56

4   Chemical reactions 58
    4.1   Basic properties of reactions 58
    4.2   The reversible linear reaction 60
    4.3   The reversible bilinear reaction 62

4.4   Connected reversible linear reactions 66
4.5   Connected reversible bilinear reactions 70
4.6   Summary 75

5   Enzyme kinetics 76
5.1   Enzyme catalysis 76
5.2   Deriving enzymatic rate laws 78
5.3   Michaelis–Menten kinetics 80
5.4   Hill kinetics for enzyme regulation 85
5.5   The symmetry model 90
5.6   Scaling dynamic descriptions 94
5.7   Summary 96

6   Open systems 97
6.1   Basic concepts 97
6.2   Reversible reaction in an open environment 100
6.3   Michaelis–Menten kinetics in an open environment 104
6.4   Summary 107

PART II. BIOLOGICAL CHARACTERISTICS

7   Orders of magnitude 111
7.1   Cellular composition and ultra-structure 111
7.2   Metabolism 116
7.3   Macromolecules 124
7.4   Cell growth and phenotypic functions 128
7.5   Summary 131

8   Stoichiometric structure 132
8.1   Bilinear biochemical reactions 132
8.2   Bilinearity leads to a tangle of cycles 134
8.3   Trafficking of high-energy phosphate bonds 137
8.4   Charging and recovering high-energy bonds 145
8.5   Summary 149

9   Regulation as elementary phenomena 150
9.1   Regulation of enzymes 150
9.2   Regulatory signals: phenomenology 152
9.3   The effects of regulation on dynamic states 153
9.4   Local regulation with Hill kinetics 156
9.5   Feedback inhibition of pathways 161
9.6   Increasing network complexity 165
9.7   Summary 169

PART III. METABOLISM

10   Glycolysis 173
10.1   Glycolysis as a system 173
10.2   The stoichiometric matrix 175

10.3   Defining the steady state 181
10.4   Simulating mass balances: biochemistry 185
10.5   Pooling: towards systems biology 189
10.6   Ratios: towards physiology 199
10.7   Assumptions 202
10.8   Summary 203

11   Coupling pathways 204
11.1   The pentose pathway 204
11.2   The combined stoichiometric matrix 210
11.3   Defining the steady state 214
11.4   Simulating the dynamic mass balances 216
11.5   Pooling: towards systems biology 218
11.6   Ratios: towards physiology 219
11.7   Summary 222

12   Building networks 224
12.1   AMP metabolism 224
12.2   Network integration 231
12.3   Whole-cell models 240
12.4   Summary 241

PART IV. MACROMOLECULES

13   Hemoglobin 245
13.1   Hemoglobin: the carrier of oxygen 245
13.2   Describing the states of hemoglobin 248
13.3   Integration with glycolysis 253
13.4   Summary 257

14   Regulated enzymes 259
14.1   Phosphofructokinase 259
14.2   The steady state 265
14.3   Integration of PFK with glycolysis 269
14.4   Summary 274

15   Epilogue 275
15.1   Building dynamic models in the omics era 275
15.2   Going forward 280

APPENDIX A. Nomenclature   285

APPENDIX B. Homework problems   288

References   306

Index   314

Preface

(Molecular) Systems biology has developed over roughly the past 10 years. Its emergence has led to the development of broad genome-wide or network-wide viewpoints of organism functions that have developed against the context of whole genome sequences. Bottom-up approaches to network reconstruction have resulted in organism-specific networks that have a direct genetic and genomic basis. Such networks are now available for a growing number of organisms.

Genome-scale networks have been used to develop constraint-based reconstruction and analysis (COBRA) procedures that treat structural properties of networks, their physiological capabilities, optimal functional states of organisms, and studies of adaptive and long-term evolution. These topics are treated in the companion book that emphasizes that while biology is dynamic, it still functions under the constraints of the topological structure of the molecular networks that underlie its functions.

Events over the time scales associated with distal causation in biology, i.e., over multiple generations, can be studied within the COBRA framework. However, analysis of proximal or immediate dynamic responses of organisms is limited. The recent development of high-throughput technologies and the availability of omics data sets has opened up an alternative approach to building large-scale models that can compute the dynamic states of biological networks. Omics-based abundance measurements (i.e., for proteins, transcripts, and metabolites) can now be mapped onto network reconstructions. In addition, functional states can be determined from fluxomic, exo-metabolomic, and various physiological data types.

The combination of omics data sets and network reconstructions allows the generation of Mass Action Stoichiometric Simulation (MASS) models. Such models can, at this point in time, be formulated for metabolism and associated enzymes and other protein molecules. MASS models will be condition specific, as they use particular data sets. In principle, MASS models can be formulated for any cellular phenomena for which reconstructions and omics data sets are available. Although the procedure is now established, some of the practical issues associated with its broad implementation will need additional experience that will call on further research in this field.

This book is focused on the process and the issues associated with the generation of MASS models. Their foundational concepts are described and they are applied to specific cases. Once the reader has mastered these concepts and gone through the details of their application to familiar cellular processes, you should be able to build MASS models for cellular phenomena of interest.

One should be aware of the fact that dynamic models have been constructed to describe biological phenomena for many decades. At the biochemical level, such models have been largely based on biophysical principles, heavily focused in particular on the use of *in-vitro*-derived rate laws. Given the scarcity of such rate laws, this approach to building kinetic models has limited the scope and size of dynamic models built in this fashion. The omics data-driven MASS procedure provides an alternative condition-dependent approach that is scalable.

This book, in a sense, brings my career full circle. My first love in graduate school was building complex dynamic models in biology based on the contents of the graduate curriculum in chemical engineering. However, as stated above, the application of these methods to biology was necessarily limited due to data availability and due to the "absolute" characteristics of biophysical models. The path through stoichiometric models from the biochemical to the genome scale based on full genome sequences, to large-scale dynamic models based on omics data sets has been an interesting one. Given the impending onslaught of genetic data and associated potential for biological variation, this field might be just in its infancy.

This text is constructed to teach how to build complex dynamic models of biochemical networks and how to simulate their responses. The material has been taught both at the undergraduate and graduate level at UC San Diego since 2008. Teaching the material at these two levels has led to the development of a set of homework problems (Appendix B) and a collection of Mathematica workbooks. It is my intent to make these available through an on-line source, initially on http://systemsbiology.ucsd.edu. I hope both will be helpful to instructors.

The path to this book has had many influences. Reich and Selkov's 1982 book, *Energy Metabolism of the Cell*, certainly contains many foundational and influential concepts. The *Color Atlas of Biochemistry* by Koolman and Roehm provides succinct representation of biochemical knowledge that has been useful in developing the material. All the computations in the text were done in Mathematica. Throughout my entire career, LATEX has been an essential resource, as it was for writing this book.

There are special thanks due to two individuals. Neema Jamshidi has been an MD/PhD student in my lab over the past 6 or 7 years. He has been a fantastic colleague and friend. He educated me about the use of Mathematica and tirelessly answered my repeated and often naive questions. He has also been a source of great intellectual stimulation and discussions. He was a major influence in completing this book. As with the companion book, Marc Abrams made the writing, preparation, editing, and production of this book possible. He supervised, coordinated, and implemented the construction of the LATEX document and the preparation of many figures in the text. Special thanks to these two gentlemen.

In addition, three PhD students in my lab were of invaluable help in getting this book to the state of completion that it has reached. Aarash Bordbar helped me with the formulation of the complicated Mathematica workbooks for Part IV of the text. In addition, he has played a notable role in developing the work flow for MASS models. Daniel Zielinski not only helped build the Mathematica workbooks for Part III, but proofread the text with his impeccable eye for detail and logical flow of material. Addiel U. de Alba Solis helped with the Mathematica workbooks for Parts I and II of the text. All three were very helpful in reviewing, correcting, and providing solutions to the homework sets given in Appendix B.

Others have helped with this text either indirectly or directly through thoughtful comments or the preparation of illustrations. For their assistance, I am grateful: Kenyon Applebee, Tom Conrad, Markus Herrgard, Joshua Lerman, Vasiliy Portnoy, Jan Schellenberger, Paolo Vicini and Michael Zager.

This book is dedicated to my parents, who enabled, allowed, supported, and encouraged me to pursue my studies and interests in integrated biological processes. Without them I would not have reached this level of professional development and would not have written this book. Kærar þakkir.

Bernhard Palsson
La Jolla, CA
April 2010

## CHAPTER 1

Introduction

Systems biology has been brought to the forefront of life-science-based research and development. The need for systems analysis is made apparent by the inability of focused studies to explain whole network, cell, or organism behavior, and the availability of component data is what is fueling and enabling the effort. This massive amount of experimental information is a reflection of the complex molecular networks that underlie cellular functions. Reconstructed networks represent a common denominator in systems biology. They are used for data interpretation, comparing organism capabilities, and as the basis for computing their functional states. The companion book [89] details the topological features and assessment of functional states of biochemical reaction networks and how these features are represented by the stoichiometric matrix. In this book, we turn our attention to the kinetic properties of the reactions that make up a network. We will focus on the formulation of dynamic simulators and how they are used to generate and study the dynamic states of biological networks.

1.1 Biological networks

Cells are made up of many chemical constituents that interact to form networks. Networks are fundamentally comprised of *nodes* (the compounds) and the *links* (chemical transformations) between them. The networks take on functional states that we wish to compute, and it is these physiological states that we observe. This text is focused on dynamic states of networks.

There are many different kinds of biological network of interest, and they can be defined in different ways. One common way of defining networks is based on a preconceived notion of what they do. Examples include metabolic, signaling, and regulatory networks; see Figure 1.1. This

1

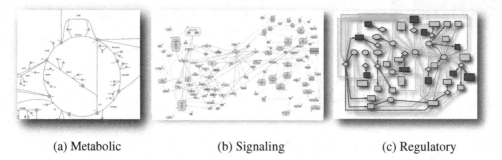

(a) Metabolic   (b) Signaling   (c) Regulatory

Figure 1.1 Three examples of networks that are defined by major function. (a) Metabolism. (b) Signaling. From Arisi *et al. BMC Neuroscience* 2006 7(Suppl 1):S6 DOI: 10.1186/1471-2202-7-S1-S6. (c) Transcriptional regulatory networks. Image courtesy of Christopher Workman, Center for Biological Sequence Analysis, Technical University of Denmark.

approach is driven by a large body of literature that has grown around a particular cellular function.

Metabolic networks Metabolism is ubiquitous in living cells and is involved in essentially all cellular functions. It has a long history – glycolysis was the first pathway elucidated in the 1930s – and is thus well known in biochemical terms. Many of the enzymes and the corresponding genes have been discovered and characterized. Consequently, the development of dynamic models for metabolism is the most advanced at the present time.

A few large-scale kinetic models of metabolic pathways and networks now exist. Genome-scale reconstructions of metabolic networks in many organisms are now available. With the current developments in metabolomics and fluxomics, there is a growing number of large-scale data sets becoming available. However, there are no genome-scale dynamic models yet available for metabolism.

Signaling networks Living cells have a large number of sensing mechanisms to measure and evaluate their environment. Bacteria have a surprising number of two-component sensing systems that inform the organism about its nutritional, physical, and biological environment. Human cells in tissues have a large number of receptor systems in their membranes to which specific ligands bind, such as growth factors or chemokines. Such signaling influences the cellular fate processes: differentiation, replication, apoptosis, and migration.

The functions of many of the signaling pathways that is initiated by a sensing event are presently known, and this knowledge is becoming more detailed. Only a handful of signaling networks are well known,

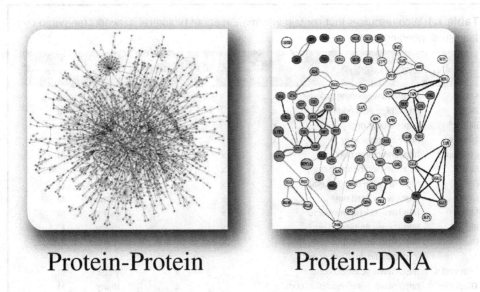

Protein-Protein   Protein-DNA

Figure 1.2 Two examples of networks that are defined by high-throughput chemical assays. Images courtesy of Markus Herrgard.

such as the JAK-STAT signaling network in lymphocytes and the Toll-like receptor system in macrophages. A growing number of dynamic models for individual signaling pathways are becoming available.

Regulatory networks There is a complex network of interactions that determine the DNA binding state of most proteins, which in turn determine whether genes are being expressed. The RNA polymerase must bind to DNA, as do transcription factors and various other proteins. The details of these chemical interactions are being worked out, but in the absence of such information, most of the network models that have been built are discrete, stochastic, and logistical in nature.

With the rapid development of experimental methods that measure expression states, the binding sites, and their occupancy, we may soon see large-scale reconstructions of transcriptional regulatory networks. Once these are available, we can begin to plan the process to build models that will describe their dynamic states.

Unbiased network definitions An alternative way to define networks is based on chemical assays. Measuring all protein–protein interactions regardless of function provides one such example; see Figure 1.2. Another example is a genome-wide measurement of the binding sites of a DNA-binding protein. This approach is driven by data-generating capabilities. It does not have an *a priori* bias about the function of molecules being examined.

Table 1.1 Web resources that contain information about biological networks (prepared by Jan Schellenberger)

		Metabolic	Protein–protein	Regulatory/signaling	Organisms	Curation[a]
KEGG	http://www.genome.jp/kegg/	x			many	C
BiGG	http://bigg.ucsd.edu/	x			many	M
BioCyc[b]	http://biocyc.org/	x		x	many	C/M
MetaCyc	http://metacyc.org/	x			many	C/M
Reactome	http://reactome.org/	x	x	x	many	M
BIND	http://www.bindingdb.org/		x		many	E/M
DIP	http://dip.doe-mbi.ucla.edu/		x		many	M
HPRD	http://www.hprd.org/		x		human	M
MINT	http://mint.bio.uniroma2.it/		x		many	M
Biogrid	http://www.thebiogrid.org/		x		many	E
UniHI	http://theoderich.fb3. mdc-berlin.de:8080/unihi/		x		human	E/M
Yeastract	http://www.yeastract.com/			x	yeast	M
TRANSFAC	http://www.gene-regulation.com			x	many	M
TRANSPATH	http://www.gene-regulation.com			x	many	M
RegulonDB	http://regulondb.ccg.unam.mx/			x	many	C/E
NetPath	http://www.netpath.org/			x	human	M

[a] M = manual/literature, C = computational, E = experimental.
[b] Links to other *Cyc databases.

Network reconstruction Metabolic networks are currently the best-characterized biological networks for which the most detailed reconstructions are available. The conceptual basis for their reconstruction has been reviewed [100], the workflows used detailed [35], and a detailed standard operating procedure (SOP) is available [117]. Some of the fundamental issues associated with the generation of dynamic models describing their functions have been articulated [52].

There is much interest in reconstructing signaling and regulatory networks in a similar way. The prospects for reconstruction of large-scale signaling networks have been discussed [49]. Given the development of new omics data types and other information, it seems likely that we will be able to obtain reliable reconstructions of these networks in the not too distant future.

Public information about pathways and networks There is a growing number of networks that underlie cellular functions that are being unraveled and reconstructed. Many publicly available sources contain this information; see Table 1.1. We wish to study the dynamic states of such networks. To do so, we need to describe them in chemical detail and incorporate thermodynamic information and formulate a mathematical model.

1.2   Why build and study models?

Mathematical modeling is practiced in various branches of science and engineering. The construction of models is a laborious and detailed task. It also involves the use of numerical and mathematical analysis, both of which are intellectually intensive and unforgiving undertakings. So why bother?

Bailey's five reasons   The purpose and utility of model building has been succinctly summarized and discussed [15]:

1. *"To organize disparate information into a coherent whole."* The information that goes into building models is often found in many different sources and the model builder has to look for these, evaluate them, and put them in context. In our case, this comes down to building data matrices (see Table 1.3) and determining conditions of interest.
2. *"To think (and calculate) logically about what components and interactions are important in a complex system."* Once the information has been gathered it can be mathematically represented in a self-consistent format. Once equations have been formulated using the information gathered and according to the laws of nature, the information can be mathematically interrogated. The interactions among the different components are evaluated and the behavior of the model is compared with experimental data.
3. *"To discover new strategies."* Once a model has been assembled and studied, it often reveals relationships among its different components that were not previously known. Such observations often lead to new experiments, or form the basis for new designs. Further, when a model fails to reproduce the functions of the process being described, it means there is either something critical missing in the model or the data that led to its formulation is inconsistent. Such an occurrence then leads to a re-examination of the information that led to the model formulation. If no logical flaw is found, the analysis of the discrepancy may lead to new experiments to try to discover the missing information.
4. *"To make important corrections to the conventional wisdom."* The properties of a model may differ from the governing thinking about

process phenomena that is inferred based on qualitative reasoning. Good models may thus lead to important new conceptual developments.

5. *"To understand the essential qualitative features."* Since a model accounts for all interactions described among its parts, it often leads to a better understanding of the whole. In the present case, such qualitative features relate to multi-scale analysis in time and an understanding of how multiple chemical events culminate in coherent physiological features.

1.3   Characterizing dynamic states

The dynamic analysis of complex reaction networks involves the tracing of time-dependent changes of concentrations and reaction fluxes over time. The concentrations typically considered are those of metabolites, proteins, or other cellular constituents. There are three key characteristics of dynamic states that we mention here right at the outset, and they are described in more detail in Section 2.1.

Time constants   Dynamic states are characterized by change in time; thus, the rate of change becomes the key consideration. The rate of change of a variable is characterized by a *time constant*. Typically, there is a broad spectrum of time constants found in biochemical reaction networks. This leads to time-scale separation, where events may be happening on the order of milliseconds all the way to hours, if not days. The determination of the spectrum of time constants is thus central to the analysis of network dynamics.

Aggregate variables   An associated issue is the identification of the biochemical, and ultimately physiological, events that are unfolding on every time scale. Once identified, one begins to form *aggregate concentration variables*, or *pooled variables*. These variables will be combinations of the original concentration variables. For example, two concentration variables may interconvert rapidly, on the order of milliseconds and thus on every time scale longer than milliseconds these two concentrations will be "connected." They can, therefore, be "pooled" together to form an aggregate variable. An example is given in Figure 1.3.

The determination of such aggregate variables becomes an intricate mathematical problem. Once solved, it allows us to determine the *dynamic structure of a network*. In other words, we move hierarchically away from the original concentration variables to increasingly interlinked aggregate

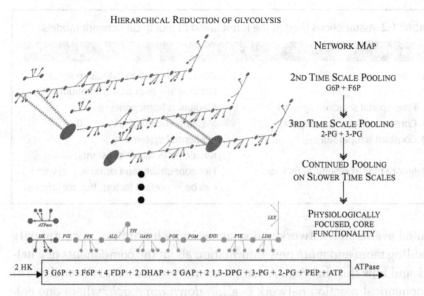

Figure 1.3 Time-scale hierarchy and the formation of aggregate variables in glycolysis. The "pooling" process culminates in the formation of one pool (shown in a box at the bottom) that is filled by hexokinase (HK) and drained by ATPase. This pool represents the inventory of high-energy phosphate bonds. From [52].

variables that ultimately culminate in the overall dynamic features of a network on slower time scales. Temporal decomposition, therefore, involves finding the time-scale spectrum of a network and determining what moves on each one of these time scales. A network can then be studied on any one of these time scales.

Transitions Complex networks can transition from one steady state (i.e., homeostatic state) to another. There are distinct types of transition that characterize the dynamic states of a network. Transitions are analyzed by *bifurcation theory*. The most common bifurcations involve the emergence of *multiple steady states*, *sustained oscillations*, and *chaotic* behavior. Such dynamic features call for a yet more sophisticated mathematical treatment. Such changes in dynamic states have been called *creative functions*, which in turn represent willful physiological changes in organism behavior. In this book, we will only encounter relatively simple types of such transition.

1.4 Formulating dynamic network models

Approach Mechanistic kinetic models based on differential equations represent a *bottom-up approach*. This means that we identify all the

Table 1.2 Assumptions used in the formulation of biological network models

Assumption	Description
(1) Continuum assumption	Do not deal with individual molecules, but treat medium as a continuum
(2) Finer spatial structure ignored	Medium is homogeneous
(3) Constant-volume assumption	V is time-invariant, $dV/dt = 0$
(4) Constant temperature	Isothermal systems
	Kinetic properties a constant
(5) Ignore physico-chemical factors	Electroneutrality and osmotic pressure can be important factors, but are ignored

detailed events in a network and systematically build it up in complexity by adding more and more new information about the components of a network and how they interact. A complementary approach to the analysis of a biochemical reaction network is a *top-down approach*, where one collects data and information about the state of the whole network at one time. This approach is not covered in this text but typically requires a Bayesian or Boolean analysis that represents causal or statistically determined relationships between network components. The bottom-up approach requires a mechanistic understanding of component interactions. Both the top-down and bottom-up approaches are useful and complementary in studying the dynamic states of networks.

Simplifying assumptions   Kinetic models are typically formulated as a set of deterministic ordinary differential equations (ODEs). There are a number of important assumptions made in such formulations that often are not fully described and delineated. Five assumptions will be discussed here (Table 1.2).

1. Using deterministic equations to model biochemistry essentially implies a "clockwork" of functionality. However, this modeling assumption needs justification. There are three principal sources of variability in biological dynamics: internal thermal noise, changes in the environment, and cell-to-cell variation. Inside cells, all components experience thermal effects that result in *random molecular motion*. This process is, of course, one of molecular diffusion, called *Brownian motion* with larger observable objects. The ODE assumption involves taking an ensemble of molecules and averaging out the stochastic effects. In cases where there are very few molecules of a particular species inside a cell or a cellular compartment, this assumption may turn out to be erroneous.

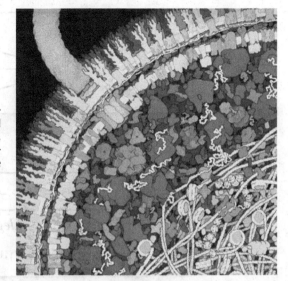

Figure 1.4 The crowded state of the intra-cellular environment. Some of the physical characteristics are viscosity (>100 × μ_{H_2O}), osmotic pressure (<150 atm), electrical gradients (≈300 000 V/cm), and a near-crystalline state. ©David S. Goodsell 1999.

2. The finer architecture of cells is also typically not considered in kinetic models. Cells are highly structured down to the 100 nm length scale and are thus not homogeneous (see Figure 1.4). Rapidly diffusing compounds, such as metabolites and ions, will distribute quickly throughout the compartment and one can justifiably consider the concentration to be relatively uniform. However, with larger molecules whose diffusion is hindered and confined, one may have to consider their spatial location. Studying and describing cellular functions of the 100 nm length scale is likely to represent an interesting topic in systems biology as it unfolds.

3. Another major assumption in most kinetic models is that of *constant volume*. Cells and cellular compartments typically have fluctuations in their volume. Treating variable volume turns out to be mathematically difficult. It is, therefore, often ignored. However, minor fluctuations in the volume of a cellular compartment may change all the concentrations in that compartment and, therefore, all kinetic and regulatory effects.

4. Temperature is typically considered to be a constant. Larger organisms have the capability to control their temperature. Small organisms have a high surface-to-volume ratio, making it hard to control heat flux at their periphery. Further, small cellular dimensions lead to rapid thermal diffusivity and a strong dependency on the thermal characteristics of the environment. Rate constants are normally a strong function of temperature, often described by the Arrhenius law. Thus, treating cells as isothermal systems is a simplification under which the kinetic properties are described by kinetic *constants*.

(a)

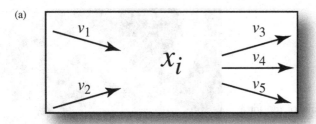

(b)

$$\frac{dx_i}{dt} = \underbrace{v_1 + v_2}_{formation} - \underbrace{v_3 - v_4 - v_5}_{degradation}$$

$$= \left\langle (1,1,-1,-1,-1), (v_1,v_2,v_3,v_4,v_5)^T \right\rangle$$

Figure 1.5 The dynamic mass balance on a single compound. (a) All the rates of formation and degradation of a compound x_i (a graphical representation called a node map). (b) The corresponding dynamic mass balance equation that simply states that the rate of change of the concentrations x_i is equal to the sum of the rates of formation minus the sum of the rates of degradation. This summation can be represented as an inner product between a row vector and the flux vector. This row vector becomes a row in the stoichiometric matrix in Eq. (1.1).

5. All cells and cellular compartments must maintain electroneutrality; therefore, the exchange of any species in and out of a compartment or a cell must also obey electroneutrality. Considering the charge of molecules and their pH dependence is yet another complicated mathematical subject and, thus, often ignored. Similarly, significant internal osmotic pressure must be balanced with that of the environment. Cells in tissues are in an isotonic environment, while single-cellular organisms and cells in plants build rigid walls to maintain their integrity.

The dynamic mass balance equations Applying these simplifying assumptions, we arrive at the dynamic mass balance equations as the starting point for modeling the dynamic states of biochemical reaction networks. The basic notion of a dynamic mass balance on a single compound, x_i, is shown in Figure 1.5.

The combination of all the dynamic mass balances for all concentrations **x** in a biochemical reaction network are given by a matrix equation:

$$\frac{d\mathbf{x}}{dt} = \mathbf{Sv(x)}, \tag{1.1}$$

where $\mathbf{S}$ is the stoichiometric matrix, $\mathbf{v}$ is the vector of reaction fluxes (v_j), and $\mathbf{x}$ is the vector of concentrations (x_i). Equation (1.1) will be the "master" equation that will be used to describe network dynamics states in this book.

Alternative views There are a number of considerations that come with the differential equation formalism described in this book and in the vast majority of the research literature on this subject matter. Perhaps the most important issue is the treatment of cells as behaving deterministically and displaying continuous changes over time. It is possible, though, that cells ultimately will be viewed as essentially a *liquid crystalline state* where transitions will be discrete from one state to the next, and not continuous.

1.5 The basic information is in a matrix format

The natural mathematical language for describing network states using dynamic mass balances is that of matrix algebra. In studying the dynamic states of networks, there are three fundamental matrices of interest: the stoichiometric, the gradient, and the Jacobian matrices.

The stoichiometric matrix The stoichiometric matrix, the properties of which are detailed in [89], represents the reaction topology of a network. Every row in this matrix represents a compound (alternative states require multiple rows) and every column represents a link between compounds, or a chemical reaction. All the entries in this matrix are *stoichiometric coefficients*. This matrix is "knowable," since it is comprised of integers that have no error associated with them. The stoichiometric matrix is *genomically derived* and thus all members of a species or a biopopulation will have the same stoichiometric matrix.

Mathematically, the stoichiometric matrix has important features. It is a *sparse matrix*, which means that few of its elements are nonzero. Typically, less than 1% of the elements of a genome-scale stoichiometric matrix are nonzero, and those nonzero elements are almost always $+1$ or -1. Occasionally there will be an entry of numerical value 2, which may represent the formation of a homodimer. The fact that all the elements of $\mathbf{S}$ are of the same order of magnitude makes it a convenient matrix to deal with from a numerical standpoint. The properties of the stoichiometric matrix have been extensively studied [89].

The gradient matrix Each link in a reaction map has kinetic properties with which it is associated. The reaction rates that describe the kinetic properties are found in the rate laws, $\mathbf{v}(\mathbf{x}; \mathbf{k})$, where the vector $\mathbf{k}$ contains

Table 1.3 Comparison of some of the attributes and properties of the stoichiometric and the gradient matrices. Adapted from [52]

Properties	S	G
Informatic	Annotated genome	Kinetic data
	Bibliomic	Metabolomics
	Comparative Genomics	Fluxomics
Physico-chemical	Chemistry	Kinetics
	Conservations	Thermodynamics
Genetic	Genomic characteristics	Genetic characteristics
	Represents a species	Represents an individual
Biological	Species differences	Individual differences
	Distal causation	Proximal causation
Mathematical	Integer entries	Real numbers
	Knowable matrix	Entries have errors
Systemic	Pool formation	Time-scale separation
	Network structure	Dynamic function
Numerical	Sparse	Sparse
	Well-conditioned	Ill-conditioned
	Non-stiff	Leads to stiffness

all the kinetic constants that appear in the rate laws. Ultimately, these properties represent time constants that tell us how quickly a link in a network will respond to the concentrations that are involved in that link. The reciprocal of these time constants is found in the gradient matrix $\mathbf{G}$, whose elements are

$$g_{ij} = \frac{\partial v_i}{\partial x_j} \quad [\text{time}^{-1}]. \tag{1.2}$$

These constants may change from one member to the next in a biopopulation, given the natural sequence diversity that exists. Therefore, the gradient matrix is a *genetically determined* matrix. Two members of the population may have a different $\mathbf{G}$ matrix. This difference is especially important in cases where mutations exist that significantly change the kinetic properties of critical steps in the network and change its dynamic structure. Such changes in the properties of a single link in a network may change the properties of the entire network.

Mathematically speaking, $\mathbf{G}$ has several challenging features. Unlike the stoichiometric matrix, its numerical values vary over many orders of magnitude. Some links have very fast response times, while others have long response times. The entries of $\mathbf{G}$ are real numbers and, therefore, are not "knowable." The values of $\mathbf{G}$ will always come with an error bar associated with the experimental method used to determine them. Sometimes only order-of-magnitude information about the numerical values of the entries in $\mathbf{G}$ is sufficient to allow us to determine the overall dynamic properties of a network. The matrix $\mathbf{G}$ has the same sparsity properties as the matrix $\mathbf{S}$.

The Jacobian matrix   The Jacobian matrix $\mathbf{J}$ is the matrix that character-izes network dynamics. It is a product of the stoichiometric matrix and the gradient matrix. The stoichiometric matrix gives us network structure and the gradient matrix gives us kinetic parameters of the links in the network. The product of these two matrices gives us the network dynamics.

The three matrices described above are thus not independent. The Jaco-bian is given by

$$\mathbf{J} = \mathbf{SG}. \tag{1.3}$$

The properties of $\mathbf{S}$ and $\mathbf{G}$ are compared in Table 1.3.

1.6   Studying dynamic models

Simulating versus analyzing dynamic models   Large-scale dynamic models represent, in an integrated form, the knowledge that was used to formulate them. The components of the network are all dynamically interrelated, and these relationships are represented in the model. The study of dynamic states falls into two main categories: *simulation* of the model in specific circumstances and *analysis* of model properties (Figure 1.6).

- *Numerically simulating dynamic models:* Simulation involves the computation of the time-dependent behavior of the concentrations of the components of a network. Obtaining such numerical data involves the specification of kinetic constants and initial conditions, followed by the calculation of numerical solutions for the stated differential equations. The dynamic interactions among network components are computed to give the changes in concentrations and flux values over time. The results are then typically graphed and studied. Simulation represents *case studies* through the definition of the specific condi-tions being considered. Dynamic simulation is described in this text and is a fairly straightforward process to implement.
- *Mathematically analyzing dynamic models:* The dynamic properties of a network can be studied by analyzing the characteristics of the model equations themselves. This approach involves the mathemati-cal study of the model. This approach allows us to formally study

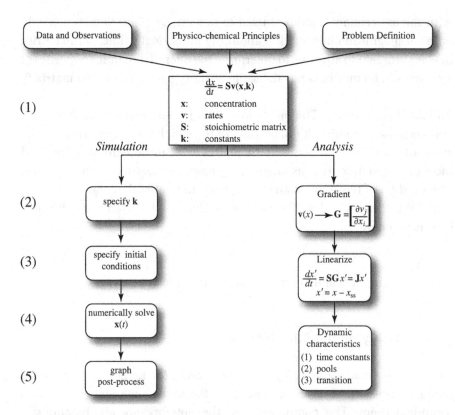

Figure 1.6 The overall process of model building to describe dynamic states of networks. The simulation approach on the left side of the diagram is covered in this text.

network properties, such as time-scale hierarchies, without using a context-specific simulation of the model describing the network. Analysis of the model properties can be performed without full specification of numerical values for parameters and initial conditions. Such studies are, therefore, not case studies, but rather give *inherent* properties of a model. Model analysis can be a mathematically and intellectually demanding process. The properties of the Jacobian matrix are central to analyzing the properties of network dynamics.

The scope and structure of this book This text will focus on the construction and study of dynamic simulation models based on the dynamic mass balances shown in Eq. (1.1). We will do this in a stepwise fashion beginning with kinetics and simulation and ending with realistic models of cellular processes (Figure 1.7).

- In Chapter 2 we review the basic concepts associated with characterizing and understanding dynamic models. Then in Part I of the book

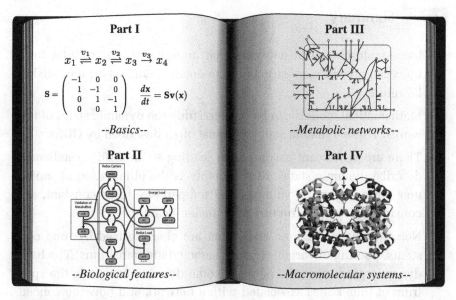

Figure 1.7 The overall organization of this text. Part I: the simulation process. Part II: general biological features of networks. Part III: basic metabolic networks. Part IV: integrated protein and small molecule networks.

we learn about the simulation procedure. We set up and examine a series of simple models to gain an understanding of how the basic concepts associated with dynamic simulation manifest themselves.

- In Part II we begin the process of looking at biologically relevant issues. We study how to deal with numerical values of the needed parameters, study the overarching effects of network structure, and learn how dynamic states are changed by regulatory mechanisms.

- In Part III we begin to build large simulators of complex biochemical reaction networks. We focus on a step-by-step construction of a dynamic simulator that describes the basic metabolic pathways in the human red blood cell. This simulator is comprised of low molecular weight compounds: the metabolites.

- The metabolites can bind to proteins and alter their properties. Thus, in Part IV, we learn how to include proteins and their bound states by focusing on the inclusion of hemoglobin and phosphofructokinase (PFK) in the red cell metabolic simulation models.

After going through the entire book the reader should understand the basics of formulating network models, how to simulate them, how to interpret their responses, and how to scale them up to biologically meaningful size, provided the data needed is available.

1.7   Summary

➤ Large-scale biological networks that underlie various cellular functions can now be reconstructed from detailed data sets and published literature.

➤ Mathematical models can be built to study the dynamic states of networks. These dynamic states are most often described by ODEs.

➤ There are significant assumptions leading to an ODE formalism to describe dynamic states. Most notably is the elimination of molecular noise, assuming volumes and temperature to be constant, and considering spatial structure to be insignificant.

➤ Networks have dynamic states that are characterized by time constants, pooling of variables, and characteristic transitions. The basic dynamic properties of a network come down to analyzing the spectrum of time scales associated with a network and how the concentrations move on these time scales.

➤ Sometimes the concentrations move in tandem at certain time scales, leading to the formation of aggregate variables. This property is key to the hierarchical decomposition of networks and the understanding of how physiological functions are formed.

➤ The data used to formulate models to describe dynamic states of networks is found in two matrices: the stoichiometric matrix, that is typically well known, and the gradient matrix, whose elements are harder to determine.

➤ Dynamic states can be studied by simulation or mathematical analysis. This text is focused on the process of dynamic simulation and its uses.

## CHAPTER 2

Basic concepts

The bottom-up analysis of dynamic states of a network is based on network topology and kinetic theory describing the links in the network. In this chapter, we provide a primer for the basic concepts of dynamic analysis of network states. We also discuss the basics of kinetic theory needed to formulate and understand detailed dynamic models of biochemical reaction networks.

2.1   Properties of dynamic states

The three key dynamic properties outlined in the introduction – time constants, aggregate variables and transitions – are detailed in this section.

Time scales A fundamental quantity in dynamic analysis is the *time constant*. A time constant is a measure of time span over which significant changes occur in a state variable. It is thus a scaling factor for time and determines where in the time-scale spectrum one needs to focus attention when dealing with a particular process or event of interest.

A general definition of a time constant is given by

$$\tau = \frac{\Delta x}{|dx/dt|_{\text{avg}}}, \tag{2.1}$$

where Δx is a characteristic change in the state variable x of interest and $|dx/dt|_{\text{avg}}$ is an estimate of the rate of change of the variable x. Notice the ratio between Δx and the average derivative has units of time, and the time constant characterizes the time span over which these changes in x occur; see Figure 2.1.

In a network, there are many time constants. In fact, there is a spectrum of time constants, $\tau_1, \tau_2, \ldots, \tau_r$, where r is the rank of the Jacobian

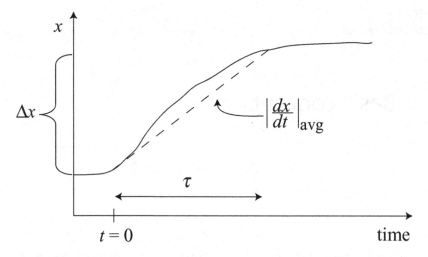

Figure 2.1 Illustration of the concept of a time constant τ and its estimation as $\tau = \Delta x/|dx/dt|_{\text{avg}}$.

matrix defining the dynamic dimensionality of the dynamic response of the network. This spectrum of time constants typically spans many orders of magnitude. The consequences of a well-separated set of time constants is a key concern in the analysis of network dynamics.

Forming aggregate variables through "pooling" One important consequence of time-scale hierarchy is the fact that we will have fast and slow events. If fast events are filtered out or ignored, then one removes a dynamic degree of freedom from the dynamic description, thus reducing the dynamic dimension of a system. Removal of a dynamic dimension leads to "coarse-graining" of the dynamic description. Reduction in dynamic dimension results in the combination, or pooling, of variables into aggregate variables.

A simple example can be obtained from upper glycolysis. The first three reactions of this pathway are

$$\text{glucose} \quad \overset{\text{HK}}{\underset{\overset{\frown}{\text{ATP ADP}}}{\rightarrow}} \quad \text{G6P} \quad \underset{\text{fast, } \tau_f}{\overset{\text{PGI}}{\rightleftharpoons}} \quad \text{F6P} \quad \overset{\text{PFK}}{\underset{\overset{\frown}{\text{ATP ADP}}}{\rightarrow}} \quad \text{FDP}. \tag{2.2}$$

This schema includes the second step in glycolysis where glucose-6-phosphate (G6P) is converted to fructose-6-phosphate (F6P) by the phosphogluco-isomerase (PGI). Isomerases are highly active enzymes and have rate constants that tend to be fast. In this case, PGI has a much faster response time than the response time of the flanking kinases in this pathway, hexokinase (HK) and PFK. If one considers a time period that is much greater than τ_f (the time constant associated with PGI), this system

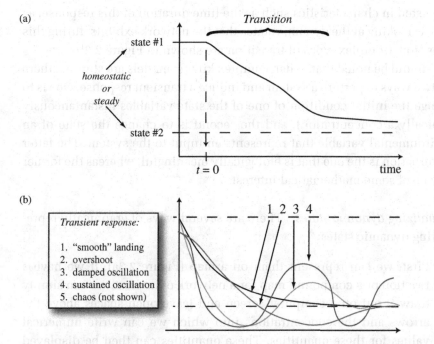

Figure 2.2 Illustration of a transition from one state to another: (a) a simple transition; (b) a more complex set of transitions.

is simplified to

$$\overset{\text{HK}}{\longrightarrow} \underset{\text{ATP}\frown\text{ADP}}{\quad} \underset{t \gg \tau_f}{\overset{\text{HP}}{\longrightarrow}} \overset{\text{PFK}}{\longrightarrow} \underset{\text{ATP}\frown\text{ADP}}{\quad}, \qquad (2.3)$$

where HP = (G6P + F6P) is the hexosephosphate pool. At a slow time scale (i.e., long compared with τ_f), the isomerase reaction has effectively equilibrated, leading to the removal of its dynamics from the network. As a result, F6P and G6P become dynamically coupled and can be considered to be a single variable. HP is an example of an aggregate variable that results from pooling G6P and F6P into a single variable. Such aggregation of variables is a consequence of time-scale hierarchy in networks. Determining how to aggregate variables into meaningful quantities becomes an important consideration in the dynamic analysis of network states. Further examples of pooling variables are given in Section 2.3.

Transitions　The dynamic analysis of a network comes down to examining its transient behavior as it moves from one state to another.

　One type of transition, or *transient response*, is illustrated in Figure 2.2a, where a system is in a homeostatic state, labeled as state #1, and is perturbed at time zero. Over some time period, as a result of the perturbation, it transitions into another homeostatic state (state #2). We are interested in characteristics such as the time duration of this response, as well as looking at the dynamic states that the network exhibits during this transition. Complex types of transition are shown in Figure 2.2b.

It should be noted that, when complex kinetic models are studied, there are two ways to perturb a system and induce a transient response. One is to change the initial condition of one of the state variables instantaneously (typically a concentration), and the second is to change the state of an environmental variable that represents an input to the system. The latter perturbation is the one that is biologically meaningful, whereas the former may be of some mathematical interest.

Visualizing dynamic states   There are several ways of graphically representing dynamic states:

- First, we can represent them on a map (Figure 2.3a). If we have a reaction or a compound map for a network of interest, we can simply draw it out on a computer screen and leave open spaces above the arrows and the concentrations into which we can write numerical values for these quantities. These quantities can then be displayed dynamically as the simulation proceeds, or by a graph showing the changes in the variable over time. This representation requires writing complex software to make such an interface.
- A second, and probably more common, way of viewing dynamic states is to simply graph the state variables x as a function of time (Figure 2.3b). Such graphs show how the variables move up and down, and on which time scales. Often, one uses a logarithmic scale for the y-axis, and that often delineates the different time constants on which a variable moves.
- A third way to represent dynamic solutions is to plot two state variables against one another in a two-dimensional plot (Figure 2.3c). This representation is known as a *phase portrait*. Plotting two variables against one another traces out a curve in this plane along which time is a parameter. At the beginning of the trajectory the time is zero, and at the end the time has gone to infinity. These phase portraits will be discussed in more detail in Chapter 3.

2.2   Primer on rate laws

The reaction rates, v_i, are described mathematically using kinetic theory. In this section, we will discuss some of the fundamental concepts of kinetic theory that lead to their formation.

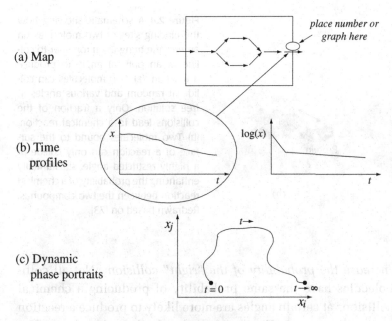

(a) Map

(b) Time profiles

(c) Dynamic phase portraits

Figure 2.3 Graphical representation of dynamic states.

Elementary reactions The fundamental events in chemical reaction networks are elementary reactions. There are two types of elemental reaction:

$$
\begin{aligned}
\text{linear} \quad & x \xrightarrow{v} \\
\text{bilinear} \quad & x_1 + x_2 \xrightarrow{v}.
\end{aligned}
\tag{2.4}
$$

A special case of a bilinear reaction is when x_1 is the same as x_2, in which case the reaction is second order.

Elementary reactions represent the irreducible events of chemical transformations, analogous to a base pair being the irreducible unit of DNA sequence. Note that rates v and concentrations x are nonnegative variables; that is:

$$
x \geq 0, \quad v \geq 0.
\tag{2.5}
$$

Mass action kinetics The fundamental assumption underlying the mathematical description of reaction rates is that they are proportional to the collision frequency of molecules taking part in a reaction. Most commonly, reactions are bilinear, where two different molecules collide to produce a chemical transformation. The probability of a collision is proportional to the concentration of a chemical species in a three-dimensional unconstrained domain. This proportionality leads to the following elementary reaction rates:

$$
\begin{aligned}
\text{linear} \quad & v = kx && \text{where the units on } k \text{ are time}^{-1} \text{ and} \\
\text{bilinear} \quad & v = kx_1x_2 && \text{where the units on } k \text{ are time}^{-1} \text{conc}^{-1}.
\end{aligned}
\tag{2.6}
$$

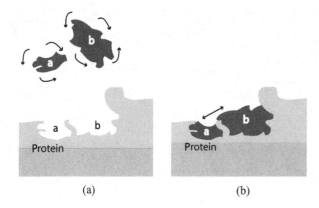

Figure 2.4 A schematic showing how the binding sites of two molecules on an enzyme bring them together to collide at an optimal angle to produce a reaction. (a) Two molecules can collide at random and various angles in free solution. Only a fraction of the collisions lead to a chemical reaction. (b) Two molecules bound to the surface of a reaction can only collide at a highly restricted angle, substantially enhancing the probability of a chemical reaction between the two compounds. Redrawn based on [73].

Enzymes increase the probability of the "right" collision   Not all collisions of molecules have the same probability of producing a chemical reaction. Collisions at certain angles are more likely to produce a reaction than others. As illustrated in Figure 2.4, molecules bound to the surface of an enzyme can be oriented to produce collisions at certain angles, thus accelerating the reaction rate. The numerical values of the rate constants are thus genetically determined as the structure of a protein is encoded in the sequence of the DNA. Sequence variation in the underlying gene in a population leads to differences amongst the individuals that make up the population. Principles of enzyme catalysis are discussed further in Section 5.1.

Generalized mass action kinetics   The reaction rates may not be proportional to the concentration in certain circumstances, and we may have what are called *power-law kinetics*. The mathematical forms of the elementary rate laws are

$$v = kx^a,$$
$$v = kx_1^a x_2^b, \tag{2.7}$$

where a and b can be greater or smaller than unity. In cases where a restricted geometry reduces the probability of collision relative to a geometrically unrestricted case, the numerical values of a and b are less than unity, and vice versa.

Combining elementary reactions   In the analysis of chemical kinetics, the elementary reactions are often combined into reaction mechanisms. Two such examples follow.

(1) *Reversible reactions:* If a chemical conversion is thermodynamically reversible, then the two opposite reactions can be combined as

$$x_1 \underset{v^-}{\overset{v^+}{\rightleftharpoons}} x_2.$$

The net rate of the reaction can then be described by the difference between the forward and reverse reactions:

$$v_{net} = v^+ - v^- = k^+ x_1 - k^- x_2, \quad K_{eq} = x_2/x_1 = k^+/k^-, \qquad (2.8)$$

where K_{eq} is the equilibrium constant for the reaction. Note that v_{net} can be positive or negative. Both k^+ and k^- have units of reciprocal time. They are thus inverses of time constants. Similarly, a net reversible bilinear reaction can be written as

$$x_1 + x_2 \underset{v^-}{\overset{v^+}{\rightleftharpoons}} x_3.$$

The net rate of the reaction can then be described by

$$v_{net} = v^+ - v^- = k^+ x_1 x_2 - k^- x_3, \quad K_{eq} = x_3/x_1 x_2 = k^+/k^-,$$

where K_{eq} is the equilibrium constant for the reaction. The units on the rate constant (k^+) for a bilinear reaction are concentration per time. Note that we can also write this equation as

$$v_{net} = k^+ x_1 x_2 - k^- x_3 = k^+(x_1 x_2 - x_3/K_{eq}).$$

This can be a convenient form, as often the K_{eq} is a known number with a thermodynamic basis, and thus only a numerical value for k^+ needs to be estimated.

(2) *Converting enzymatic reaction mechanisms into rate laws:* Often, more complex combinations of elementary reactions are analyzed. The classical irreversible Michaelis–Menten mechanism is comprised of three elementary reactions:

$$\begin{array}{ccc} v_1 = k_1 se & v_2 = k_2 x & \\ S + E \underset{v_{-1} = k_{-1}x}{\rightleftharpoons} X & \longrightarrow & P + E, \end{array}$$

where a substrate S binds to an enzyme to form a complex X that can break down to generate the product P. The concentrations of the corresponding chemical species are denoted with the same lower case letter; i.e., $e = [E]$, etc. This reaction mechanism has two conservation quantities associated with it: one on the enzyme $e_{tot} = e + x$ and one on the substrate $s_{tot} = s + x + p$.

A quasi-steady-state assumption (QSSA), $dx/dt = 0$, is then applied to generate the classical rate law

$$\frac{ds}{dt} = \frac{-v_{m}s}{K_{m} + s} \qquad (2.9)$$

that describes the kinetics of this reaction mechanism. This expression is the best-known rate equation in enzyme kinetics. It has two parameters: the maximal reaction rate v_{m} and the Michaelis–Menten constant $K_{m} = (k_{-1} + k_{2})/k_{1}$. The use and applicability of kinetic assumptions to deriving rate laws for enzymatic reaction mechanisms is discussed in detail in Chapter 5.

It should be noted that the elimination of the elementary rates through the use of the simplifying kinetic assumptions *fundamentally changes* the mathematical nature of the dynamic description from that of bilinear equations to that of hyperbolic equations (i.e., Eq. (2.9)) and, more generally, to ratios of polynomial functions.

Pseudo-first-order rate constants (PERCs)   The effects of temperature, pH, enzyme concentrations, and other factors that influence the kinetics can often be accounted for in a condition-specific numerical value of a constant that looks like a regular elementary rate constant, as in Eq. (2.4). The advantage of having such constants is that it simplifies the network dynamic analysis. The disadvantage is that dynamic descriptions based on PERCs are condition specific. This issue is discussed in Parts III and IV of the book.

The mass action ratio (Γ)   The equilibrium relationship among reactants and products of a chemical will be familiar to the reader. For example, the equilibrium relationship for the PGI reaction (Eq. (2.8)) is

$$K_{eq} = \frac{[F6P]_{eq}}{[G6P]_{eq}}. \qquad (2.10)$$

This relationship is observed in a closed system after the reaction is allowed to proceed to equilibrium over a long time, $t \to \infty$ (which in practice has a meaning relative to the time constant of the reaction, $t \gg \tau_{f}$).

However, in a cell, as shown in Eq. (2.2), the PGI reactions operate in an "open" environment, i.e., G6P is being produced and F6P is being consumed. The reaction reaches a steady state in a cell that will have concentration values that are different from the equilibrium value. The *mass action ratio* for open systems, defined to be analogous to the equilibrium constant, is

$$\Gamma = \frac{[F6P]_{ss}}{[G6P]_{ss}}. \qquad (2.11)$$

The mass action ratio is denoted by Γ in the literature.

"Distance" from equilibrium   The numerical value of the ratio Γ/K_{eq} relative to unity can be used as a measure of how far a reaction is from equilibrium in a cell. Fast reversible reactions tend to be close to equilibrium in an open system. For instance, the net reaction rate for a reversible bilinear reaction (Eq. (2.2)) can be written as

$$v_{net} = k^+ x_1 x_2 - k^- x_3 = k^+ x_1 x_2 (1 - \Gamma/K_{eq}).$$

If the reaction is "fast," then $(k^+ x_1 x_2)$ is a "large" number and thus $(1 - \Gamma/K_{eq})$ tends to be a "small" number, since the net reaction rate is balanced relative to other reactions in the network.

Recap   These basic considerations of reaction rates and enzyme kinetic rate laws are described in much more detail in other standard sources, e.g., [113]. In this text, we are not so concerned about the details of the mathematical form of the rate laws, but rather with the order of magnitude of the rate constants and how they influence the properties of the dynamic response.

2.3   More on aggregate variables

Pools, or aggregate variables, form as a result of well-separated time constants. Such pools can form in a hierarchical fashion. Aggregate variables can be physiologically significant, such as the total inventory of high-energy phosphate bonds, or the total inventory of particular types of redox equivalents. These important concepts are perhaps best illustrated through a simple example that should be considered a primer on a rather important and intricate subject matter. Formation of aggregate variables in complex models is seen throughout Parts III and IV of this text.

Distribution of high-energy phosphate among the adenylate phosphates   In Figure 2.5 we show the skeleton structure of the transfer of high-energy phosphate bonds among the adenylates. In this figure we denote the use of ATP by v_1 and the synthesis of ATP from ADP by v_2. v_5 and v_{-5} denote the reaction rates of adenylate kinase that distributes the high-energy phosphate bonds among ATP, ADP, and AMP through the reaction

$$2\,\text{ADP} \rightleftharpoons \text{ATP} + \text{AMP}. \tag{2.12}$$

Finally, the synthesis of AMP and its degradation are denoted by v_3 and v_4 respectively. The dynamic mass balance equations that describe this

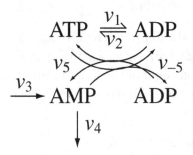

Figure 2.5 The chemical transformations involved in the distribution of high-energy phosphate bonds among adenosines.

schema are

$$\frac{d\,\text{ATP}}{dt} = -v_1 + v_2 + v_{5,\text{net}},$$

$$\frac{d\,\text{ADP}}{dt} = v_1 - v_2 - 2v_{5,\text{net}}, \qquad (2.13)$$

$$\frac{d\,\text{AMP}}{dt} = v_3 - v_4 + v_{5,\text{net}}.$$

The responsiveness of these reactions falls into three categories: $v_{5,\text{net}}$ $(= v_5 - v_{-5})$ is a *fast* reversible reaction, v_1 and v_2 have *intermediate* time scales, and the kinetics of v_3 and v_4 are *slow* and have large time constants associated with them. Based on this time-scale decomposition, we can combine the three concentrations so that they lead to the elimination of the reactions of a particular response time category on the right-hand side of Eq. (2.13). These combinations are as follows.

- First, we can eliminate all but the slow reactions by forming the sum of the adenosine phosphates:

$$\frac{d}{dt}(\text{ATP} + \text{ADP} + \text{AMP}) = v_3 - v_4 \quad \text{(slow)}. \qquad (2.14)$$

  The only reaction rates that appear on the right-hand side of the equation are v_3 and v_4, which are the slowest reactions in the system. Thus, the summation of ATP, ADP, and AMP is a pool or aggregate variable that is expected to exhibit the slowest dynamics in the system.
- The second pooled variable of interest is the summation of 2ATP and ADP that represents the total number of high-energy phosphate bonds found in the system at any given point in time:

$$\frac{d}{dt}(2\text{ATP} + \text{ADP}) = -v_1 + v_2 \quad \text{(intermediate)}. \qquad (2.15)$$

This aggregate variable is only moved by the reaction rates of intermediate response times, those of v_1 and v_2.

- The third aggregate variable we can form is the sum of the energy-carrying nucleotides, which are

$$\frac{d}{dt}(\text{ATP} + \text{ADP}) = -v_{5,\text{net}} \quad \text{(fast)}. \tag{2.16}$$

This summation will be the fastest aggregate variable in the system.

Notice that by combining the concentrations in certain ways, we define aggregate variables that may move on distinct time scales in the simple model system, and, in addition, we can interpret these variables in terms of their metabolic physiological significance. However, in general, time-scale decomposition is more complex, as the concentrations that influence the rate laws may move on many time scales and the arguments in the rate law functions must be pooled as well.

Using ratios of aggregate variables to describe metabolic physiology We can define an aggregate variable that represents the *capacity* to carry high-energy phosphate bonds. That simply is the summation of ATP + ADP + AMP. This number multiplied by 2 would be the total number of high-energy phosphate bonds that can be stored in this system. The second variable that we can define here would be the *occupancy* of that capacity, 2ATP + ADP, which is simply an enumeration of how much of that capacity is occupied by high-energy phosphate bonds. Notice that the occupancy variable has a conjugate pair, which would be the vacancy variable. The ratio of these two aggregate variables forms a charge

$$\text{charge} = \frac{\text{occupancy}}{\text{capacity}} \tag{2.17}$$

called the *energy charge*, given by

$$\text{E.C.} = \frac{2\text{ATP} + \text{ADP}}{2(\text{ATP} + \text{ADP} + \text{AMP})}, \tag{2.18}$$

which is a variable that varies between 0 and 1. This quantity is the *energy charge* defined by Daniel Atkinson [13]. In cells, the typical numerical range for this variable when measured is 0.80–0.90.

In a similar way, one can define other redox charges. For instance, the *catabolic redox charge* on the NADH carrier can be defined as

$$\text{C.R.C.} = \frac{\text{NADH}}{\text{NADH} + \text{NAD}}, \tag{2.19}$$

which simply is the fraction of the NAD pool that is in the reduced form of NADH. It typically has a low numerical value in cells, i.e., about 0.001–0.0025; therefore, this pool is typically discharged by passing the redox

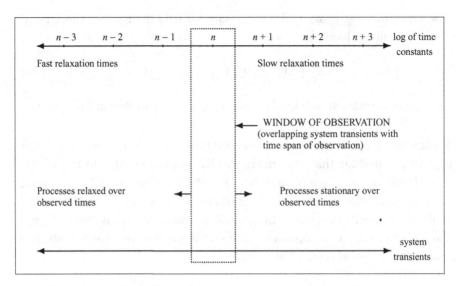

Figure 2.6 Schematic illustration of network transients that overlap with the time span of observation. $n, n+1,\ldots$ represent the decadic order of time constants.

potential to the electron transfer system (ETS). The *anabolic redox charge*

$$\text{A.R.C.} = \frac{\text{NADPH}}{\text{NADPH} + \text{NADP}},\qquad(2.20)$$

in contrast, tends to be in the range of 0.5 or higher, and thus this pool is charged and ready to drive biosynthetic reactions. Therefore, pooling variables together based on a time-scale hierarchy and chemical characteristics can lead to aggregate variables that are physiologically meaningful.

In Chapter 8 we further explore these fundamental concepts of time-scale hierarchy. They are then used in Parts III and IV in interpreting the dynamic states of realistic biological networks.

2.4   Time-scale decomposition

Reduction in dimensionality   As illustrated by the examples given in the previous section, most biochemical reaction networks are characterized by many time constants. Typically, these time constants are of very different orders of magnitude. The hierarchy of time constants can be represented by the time axis; Figure 2.6. Fast transients are characterized by the processes at the extreme left and slow transients at the extreme right. The process time scale, i.e., the time scale of interest, can be represented by a *window of observation* on this time axis. Typically, we have three principal ranges

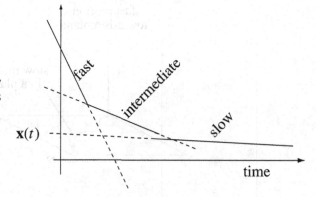

Figure 2.7 A schematic of a decay comprised of three dynamic modes with well-separated time constants.

of time constants of interest if we want to focus on a limited set of events taking place in a network. We can thus decompose the system response in time. To characterize network dynamics completely we would have to study all the time constants.

Three principal time constants One can readily conceptualize this by looking at a three-dimensional linear system where the first time constant represents the fast motion, the second represents the time scale of interest, and the third is a slow motion; see Figure 2.7. The general solution to a three-dimensional linear system is

$$
\begin{aligned}
\mathbf{x}(t) = \quad & \mathbf{v}_1 \langle \mathbf{u}_1, \mathbf{x}_0 \rangle \exp(\lambda_1 t) \quad \text{fast} \\
& + \mathbf{v}_2 \langle \mathbf{u}_2, \mathbf{x}_0 \rangle \exp(\lambda_2 t) \quad \text{intermediate} \\
& + \mathbf{v}_3 \langle \mathbf{u}_3, \mathbf{x}_0 \rangle \exp(\lambda_3 t) \quad \text{slow},
\end{aligned}
\tag{2.21}
$$

where $\mathbf{v}_i$ are the *eigenvectors*, $\mathbf{u}_i$ are the *eigenrows*, and λ_i are the *eigenvalues* of the Jacobian matrix. The eigenvalues are negative reciprocals of time constants.

The terms that have time constants faster than the observed window can be eliminated from the dynamic description, as these terms are small. However, the mechanisms which have transients slower than the observed time exhibit high "inertia" and hardly move from their initial state and can be considered constants.

Example: Three-dimensional motion simplifying to a two-dimensional motion Figure 2.8 illustrates a three-dimensional space where there is rapid motion into a slow two-dimensional subspace. The motion in the slow subspace is spanned by two "slow" eigenvectors, whereas the fast motion is in the direction of the "fast" eigenvector.

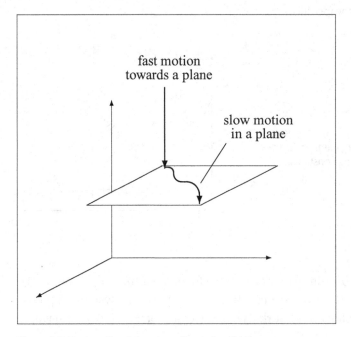

Figure 2.8 Fast motion into a two-dimensional subspace.

(a)

Network map

(b)

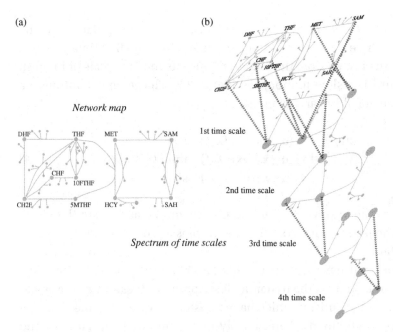

Figure 2.9 Multiple time scales in a metabolic network and the process of pool formation. This figure represents human folate metabolism. (a) A map of the folate network. (b) An illustration of progressive pool formation. Beyond the first time scale, pools form between CHF and CH2F, between 5MTHF, 10FTHF, and SAM, and between MET and SAH (these are abbreviations for the long, full names of these metabolites). DHF and THF form a pool beyond the second time scale. Beyond the third time scale, CH2F/CHF join the 5MTHF/10FTHF/SAM pool. Beyond the fourth time scale, HCY joins the MET/SAH pool. Ultimately, on time scales on the order of a minute and slower, interactions between the pools of folate carriers and methionine metabolites interact. Courtesy of Neema Jamshidi [53].

Multiple time scales In reality there are many more than three time scales in a realistic network. In metabolic systems there are typically many time scales and a hierarchical formation of pools; Figure 2.9. The formation of such hierarchies will be discussed in Parts III and IV of the text.

2.5 Network structure versus dynamics

The stoichiometric matrix represents the topological structure of the network, and this structure has significant implications with respect to what dynamic states a network can take. Its null spaces give us information about pathways and pools. It also determines the structural features of the gradient matrix. Network topology can have a dominant effect on network dynamics.

The null spaces of the stoichiometric matrix Any matrix has a right and a left null space. The right null space, normally called just the null space, is defined by all vectors that give zero when post-multiplying that matrix:

$$\mathbf{S}\mathbf{v} = 0. \tag{2.22}$$

The null space thus contains all the steady-state flux solutions for the network. The null space can be spanned by a set of basis vectors that are pathway vectors [89].

The left null space is defined by all vectors that give zero when pre-multiplying that matrix:

$$\mathbf{l}\mathbf{S} = 0. \tag{2.23}$$

These vectors $\mathbf{l}$ correspond to pools that are always conserved at all time scales. We will call them *time invariants*. Throughout the book we will look at these properties of the stoichiometric matrices that describe the networks being studied.

The structure of the gradient matrix We will now examine some of the properties of $\mathbf{G}$. If a compound x_i participates in reaction v_j, then the entry s_{ij} is nonzero. Thus, a net reaction

$$x_i + x_{i+1} \overset{v_j}{\rightleftharpoons} x_{i+2} \tag{2.24}$$

with a net reaction rate

$$v_j = v_j^+ - v_j^- \tag{2.25}$$

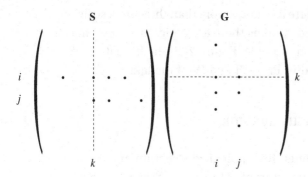

Figure 2.10 A schematic showing how the structures of **S** and **G** form matrices that have nonzero elements in the same location if one of these matrices is transposed. The columns of **S** and the rows of **G** have similar but not identical vectors in an n-dimensional space. Note that this similarity only holds once the two opposing elementary reactions have been combined into a net reaction.

generates three nonzero entries in **S**: $s_{i,j}$, $s_{i+1,j}$, and $s_{i+2,j}$. Since compounds x_i, x_{i+1}, and x_{i+2} influence reaction v_j, they will also generate nonzero elements in **G**; see Figure 2.10. Thus, nonzero elements generated by the reactions are

$$g_{j,i} = \frac{\partial v_j}{\partial x_i}, \quad g_{j,i+1} = \frac{\partial v_j}{\partial x_{i+1}}, \quad \text{and} \quad g_{j,i+2} = \frac{\partial v_j}{\partial x_{i+2}}. \tag{2.26}$$

In general, every reaction in a network is a reversible reaction. Hence, we have the the following relationships between the elements of **S** and **G**:[1]

$$\begin{aligned}
\text{if} \quad s_{i,j} &= 0 \quad \text{then} \quad g_{j,i} = 0; \\
\text{if} \quad s_{i,j} &\neq 0 \quad \text{then} \quad g_{j,i} \neq 0; \\
\text{if} \quad s_{i,j} &> 0 \quad \text{then} \quad g_{j,i} < 0; \\
\text{if} \quad s_{i,j} &< 0 \quad \text{then} \quad g_{j,i} > 0.
\end{aligned}$$

It can thus be seen that

$$-\mathbf{G}^{\mathrm{T}} \sim \mathbf{S}, \tag{2.27}$$

in the sense that both will have nonzero elements in the same location. These elements will have opposite signs.

Stoichiometric autocatalysis The fundamental structure of most catabolic pathways in a cell is such that a compound is imported into a cell and then some property stored on cofactors is transferred to the compound and the molecule is thus "charged" with this property. This charged form is then degraded into a waste product that is secreted from the cell. During that degradation process, the property that the molecule was charged with is re-extracted from the compound, often in larger quantities than was used in the initial charging of the compound. This pathway structure is

---

[1] For the rare cases where a reaction is effectively irreversible, an element in **G** can become very small, but in principle finite.

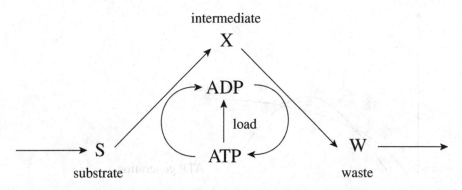

Figure 2.11 The prototypic pathway structure for degradation of a carbon substrate.

the cellular equivalent of "it takes money to make money," and its basic network structure is in Figure 2.11.

This figure illustrates the import of a substrate S to a cell. It is charged with high-energy phosphate bonds to form an intermediate X. X is then subsequently degraded to a waste product W that is secreted. In the degradation process, ATP is recouped in a larger quantity than was used in the charging process. This means that there is a net production of ATP in the two steps, and that difference can be used to drive various load functions on metabolism.

The consequence of this schema is basically *stoichiometric autocatalysis* that can lead to multiple steady states. The rate of formation of ATP from this schema as balanced by the load parameters is illustrated in Figure 2.12. This figure shows that the ATP generation is zero if all the adenosine phosphates are in the form of ATP because there is no ADP to drive the conversion of X to W. The ATP generation is also zero if there is no ATP available, because S cannot be charged to form X. The curve in between ATP = 0 and ATP = ATP_{max} will be positive. The ATP load, or use rate, will be a curve that grows with ATP concentration and is sketched here as a hyperbolic function. As shown, there are three intersections in this curve, with the upper stable steady state being the physiological state of this system. This system can thus have multiple steady states, and this property is a consequence of the topological structure of this reaction network.

Network structure The three topics discussed in this section show that the stoichiometric matrix has a dominant effect on integrated network functions and sets constraints on the dynamic states that a network can achieve. The numerical values of the elements of the gradient matrix determine which of these states are chosen.

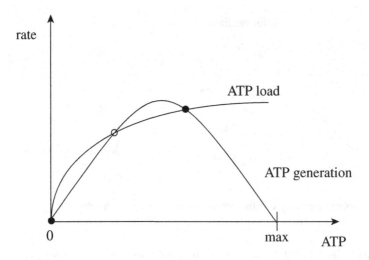

Figure 2.12 The qualitative shape of the rate of ATP generation and use in the reaction schema shown in Figure 2.11. Closed circles are dynamically stable states, whereas the open circle represents a dynamically unstable state.

2.6   Physico-chemical effects

Molecules have other physico-chemical properties besides the collision rates that are used in kinetic theory. They also have osmotic properties and are electrically charged. Both of these features influence dynamic descriptions of biochemical reaction networks.

The constant-volume assumption   Most systems that we identify in systems biology correspond to some biological entity. Such entities may be an organelle, like the nucleus or the mitochondria, or it may be the whole cell, as illustrated in Figure 2.13.

A compound x_i, internal to the system, has a mass balance on the total amount per cell. We denote this quantity with an M_i. M_i is a product of the volume per cell V and the concentration of the compound x_i, which is amount per volume:

$$M_i = V x_i. \tag{2.28}$$

The time derivative of the amount per cell is given by

$$\frac{dM_i}{dt} = \frac{d}{dt}(V x_i) = V \frac{dx_i}{dt} + x_i \frac{dV}{dt}. \tag{2.29}$$

The time change of the amount M_i per cell is thus dependent on two dynamic variables. One is dx_i/dt, which is the time change in the

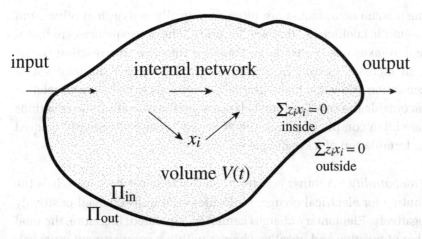

Figure 2.13 An illustration of a "system" with a defined boundary, inputs and outputs, and an internal network of reactions. The volume V of the system may change over time. Π denotes osmotic pressure; see Eq. (2.32).

concentration of x_i, and the second is dV/dt, which is the change in volume with time. The volume is typically taken to be time invariant; therefore, the term dV/dt is equal to zero and, therefore, results in a system that is of *constant volume*. In this case

$$\frac{dx_i}{dt} = \frac{1}{V}\frac{dM_i}{dt}. \tag{2.30}$$

This constant-volume assumption (recall Table 1.2) needs to be carefully scrutinized when one builds kinetic models, since volumes of cellular compartments tend to fluctuate and such fluctuations can be very important. Very few kinetic models in the current literature account for volume variation because it is mathematically challenging and numerically difficult to deal with. A few kinetic models have appeared, however, that do take volume fluctuations into account [61, 65].

Osmotic balance Molecules come with osmotic pressure, electrical charge, and other properties, all of which impact the dynamic states of networks. For instance, in cells that do not have rigid walls, the osmotic pressure has to be balanced inside (Π_{in}) and outside (Π_{out}) of the cell (Figure 2.13), i.e.,

$$\Pi_{in} = \Pi_{out}. \tag{2.31}$$

At first approximation, osmotic pressure is proportional to the total solute concentration,

$$\Pi = RT \sum_i x_i, \tag{2.32}$$

although some compounds are more osmotically active than others and have osmotic coefficients that are not unity. The consequences are that if a reaction takes one molecule and splits it into two, the reaction comes with an increase in osmotic pressure that will impact the total solute concentration allowable inside the cell, as it needs to be balanced relative to that outside the cell. Osmotic balance equations are algebraic equations that are often complicated and, therefore, are often conveniently ignored in the formulation of a kinetic model.

Electroneutrality   Another constraint on dynamic network models is the accounting for electrical charge. Molecules tend to be charged positively or negatively. Elementary charges cannot be separated; therefore, the total number of positive and negative charges within a compartment must balance. Any import and export in and out of a compartment of a charged species has to be counterbalanced by the equivalent number of molecules of the opposite charge crossing the membrane. Typically, bilipid membranes are impermeable to cations, but permeable to anions. For instance, the deliberate displacement of sodium and potassium by the ATP-driven sodium–potassium pump is typically balanced by chloride ions migrating in and out of a cell or a compartment leading to a state of electroneutrality both inside and outside the cell. The equations that describe electroneutrality are basically a summation of the charge z_i of a molecule multiplied by its concentration.

$$\sum_i z_i x_i = 0; \tag{2.33}$$

and such terms are summed up over all the species in a compartment. That sum has to add up to zero to maintain electroneutrality. Since that summation includes concentrations of species, it represents an algebraic equation that is a constraint on the allowable concentration states of a network.

2.7   Summary

➤ Time constants are key quantities in dynamic analysis. Large biochemical reaction networks typically have a broad spectrum of time constants.

➤ Well-separated time constants lead to pooling of variables to form aggregates. Aggregate variables represent a coarse-grained (i.e., lower dimensional) view of network dynamics and can lead to physiologically meaningful variables.

➤ Elementary reactions and mass action kinetics are the irreducible events in dynamic descriptions of networks. Elementary reactions are often combined into reaction mechanisms from which rate laws are derived using simplifying assumptions.

➤ Network structure has an overarching effect on network dynamics. Certain physico-chemical effects can as well. Thus, topological analysis is useful, and so is a careful examination of the assumptions (recall Table 1.2) that underlie the dynamic mass balances (Eq. (1.1)) for the system being modeled and simulated.

Simulation of dynamic states

In this part of the text we will describe the processes with which dynamic states of networks are simulated. We will first introduce the foundation of the simulation procedure. Setting up simulations involves setting up the dynamic mass balance equations, determining values for the kinetic parameters, and setting the initial conditions. The last item is important, as it emphasizes the condition-specific nature of simulation. A simulation is a *case study*. We then go though a series of chapters of increasingly complex systems to simulate, starting with simple chemical mechanisms, to simple biochemical reaction mechanisms, to open systems. Along the way we illustrate the use of simulation and elucidate key concepts that characterize dynamic states of networks.

Simulation of dynamic states

## CHAPTER 3

Dynamic simulation: the basic procedure

Once a set of dynamic mass balance equations has been formulated, they can be numerically solved, and thus the behavior of a network can be simulated in response to environmental and genetic changes. Simulation results can be obtained using a number of different software packages. Dynamic simulation generates the time-dependent behavior of the concentrations, i.e., $\mathbf{x}(t)$. This solution can be obtained in response to several different types of perturbation and the results graphically displayed. The basic principles and procedures associated with dynamic simulation are covered in this chapter. The following three chapters then apply the simulation process to a set of simple but progressively more complex and relevant examples.

3.1 Numerical solutions

Network dynamics are described by a set of ODEs: the dynamic mass balance equations; see Eq. (1.1). To obtain the dynamic solutions, we need three things: first, the equations themselves; second, the numerical values for the kinetic constants that are in the equations; and third, the initial conditions and parameters that are being perturbed. We describe each briefly.

1. To formulate the mass balances we have to specify the system boundary, the fluxes in and out of the system, and the reactions that take place in the network. From the set of reactions that are taking place, a stoichiometric matrix is formed. This matrix is then put into Eq. (1.1). One can multiply out the individual dynamic mass balances, as was done in Eq. (2.13) for the adenosine phosphate network, to prevent a large number of numerical operations that involve multiplication of

reaction rates by the zero elements in **S**. The reaction rate laws for the reactions are then identified and substituted into the equations. Typically, one would use elementary kinetics, as shown in Eq. (2.6), or apply more complex rate laws if they are appropriate and available. This process leads to the definition of the dynamic mass balances.

2. The numerical values for the kinetic parameters in the rate laws have to be specified, as do any imposed fluxes across the system boundary. Obtaining numerical values for the kinetic constants is typically difficult. They are put into a parameter vector designated by **k**. In select cases, detailed kinetic characterization has been carried out. More often, though, one only knows these values approximately. It is important to make sure that all units are consistent throughout the equations and that the numerical values used are in the appropriate units.

3. With the equations and numerical values for the kinetic constants specified (**k**), we can simulate the responses of the network that they represent. To do so, we have to set initial conditions (x_0). This leads to the numerical solution of

$$\frac{d\mathbf{x}}{dt} = \mathbf{Sv(x; k)}, \quad \mathbf{x}(t = 0) = \mathbf{x_0}. \tag{3.1}$$

There are three conditions that are typically considered.

- First, the initial conditions for the concentrations are set, and the motion of the network into a steady state (open system) or equilibrium (closed system) is simulated. This scenario is typically physiologically unrealistic, since individual concentrations cannot simply change by themselves in a living cell.

- Second, a change in an input flux is imposed on a network that is in a steady state. This scenario can be used to simulate the response of a cell to a change in its environment.

- Third, a change in a kinetic parameter is implemented at the initial time. The initial concentrations are typically set at the steady-state values with the nominal value of the parameter. The equations are then simulated to a long time to obtain the steady-state values that correspond to the altered kinetic parameters. These are set as the initial conditions when examining the responses of the system with the altered kinetic properties.

4. Once the solution has been obtained it can be graphically displayed and the results analyzed. There are several ways to accomplish this step, as detailed in the next two sections. The analysis of the results can lead to post-processing of the output to form an alternative set of dynamic variables.

Table 3.1 Available software for dynamic simulation. Assembled by Neema Jamshidi

Name	Distribution	Website
AUTO	Free	http://indy.cs.concordia.ca/auto/
Berkeley Madonna	Commercial	http://www.berkeleymadonna.com/
Cell/Designer/SBML ODE Solver	Free	http://www.systems-biology.org/002/
CONTENT	Free	http://www.computeralgebra.nl/systemsoverview/special/ diffeqns/content/gcbody.html
Copasi	Both	http://www.copasi.org/tiki-index.php
E-Cell	Free	http://www.e-cell.org/
Gepasi	Free	http://www.gepasi.org/
JACOBIAN	Commercial	http://numericatech.com/jacobian.htm
JSim	Free	http://www.physiome.org/jsim/
KinSolver	Free	http://lsdis.cs.uga.edu/~aleman/kinsolver/
Mathematica	Commercial	http://www.wolfram.com
MATLAB	Commercial	http://www.mathworks.com
PySCeS	Free	http://pysces.sourceforge.net/
SciLab	Free	http://www.scilab.org
VCell (Virtual Cell)	Free	http://www.vcell.org
WebCell	Free	http://webcell.org
XPPAUT	Free	http://www.math.pitt.edu/~bard/xpp/xpp.html

The simulation is implemented using a numerical solver. Currently, such implementation is carried out using standard and readily available software, such as Mathematica or MATLAB. Specialized simulation packages are also available (Table 3.1). After the simulation is set up and the conditions specified, the software computes the concentrations as a function of time. The output is a file that contains the numerical values of the concentrations at a series of time points (Figure 3.1). This set of numbers is typically displayed graphically and/or used for subsequent computations.

3.2 Graphically displaying the solution

The simulation procedure described in the previous section results in a file that contains the concentrations as a function of time (Figure 3.1). These results are graphically displayed, typically in two ways: by plotting the concentrations as a function of time, or by plotting two concentrations against one another with time as a parameter along the trajectory.

Before describing these methods, we observe certain fundamental aspects of the equations that we are solving. The dynamic mass balances

t	x_1	x_2	$\cdots$	x_m
t_1	$x_1(t_1)$	$x_2(t_1)$		$x_m(t_1)$
t_2	$x_1(t_2)$	$x_2(t_2)$		$x_m(t_2)$
t_3	$x_1(t_3)$	$x_2(t_3)$		$x_m(t_3)$
$\vdots$	$\vdots$			$\vdots$
t_n	$x_1(t_n)$	$x_2(t_n)$		$x_m(t_n)$
t_{n+1}	$x_1(t_{n+1})$	$x_2(t_{n+1})$		$x_m(t_{n+1})$
t_{n+2}	$x_1(t_{n+2})$	$x_2(t_{n+2})$	$\cdots$	$x_m(t_{n+2})$

Figure 3.1 The fundamental structure of the file $\mathbf{x}(t)$ that results from a numerical simulation. The two vertical bars show the list of values that would be used to compute $\sigma_{12}(2)$ (see Eq. (3.8)); that is, the correlation between x_1 and x_2 with a time lag of 2.

can be expanded as

$$\frac{d\mathbf{x}}{dt} = \mathbf{S}\mathbf{v}(\mathbf{x}) = \sum \mathbf{s}_i v_i(\mathbf{x}). \tag{3.2}$$

In other words, the time derivatives are linear combinations of the reaction vectors ($\mathbf{s}_i$) weighted by the reaction rates, which in turn change with the concentrations that are time varying. Thus, the motions are linear combinations of the directions specified by $\mathbf{s}_i$. This characteristic is important because, if the v_i have different time constants, the motion can be decomposed in time along these reaction vectors.

Time profiles The simulation results in a file that contains the vector $\mathbf{x}(t)$ and the time points at which the numerical values for the concentrations are given. These time points can be specified by the user or are automatically generated by the solver used. Typically, the user specifies the initial time, the final time, and sometimes the time increment between the time points where the simulator stores the computed concentration values in the file. The results can then be displayed graphically depending on a few features of the solution. Some of these are shown in Figure 3.2 and are now briefly described:

- Figure 3.2a: the most common way to display a dynamic solution is to plot the concentration as a function of time.
- Figure 3.2b: if there are many concentration variables they are often displayed on the same graph.
- Figure 3.2c: in many cases there are different response times and one plots multiple time profiles where the x-axis on each plot is scaled to

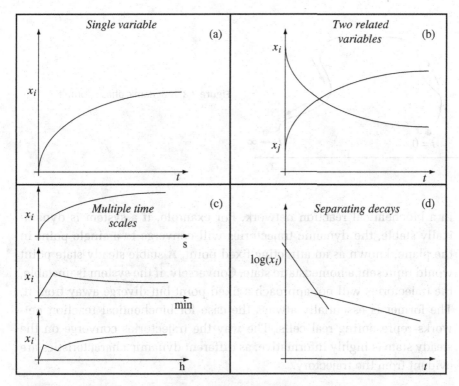

Figure 3.2 Graphing concentrations over time. (a) A single concentration shown as a function of time. (b) Many concentrations shown as a function of time. (c) A single concentration shown as a function of time separately on different time scales. (d) The logarithm of a single concentration shown as a function of time to distinguish the decay on different time scales.

    a particular response time. Alternatively, one can use a logarithmic scale for time.

- Figure 3.2d: if a variable moves on many time scales, changing over many orders of magnitude, the y-axis is often displayed on a logarithmic scale.

The solution can thus be displayed in different ways depending on the characteristics of the time profiles. One normally plays with these representations to get an understanding of the responses of the network that they have formulated and to represent the features in which one is interested.

Dynamic phase portraits Dynamic phase portraits represent trajectories formed when two concentrations are plotted against each other, parameterized with respect to time (Figure 3.3). The dynamic trajectories in the diagram move from an initial state to a final state. Analysis of these trajectories can point to key dynamic relationships between compounds

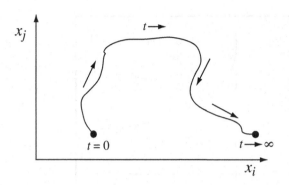

Figure 3.3 A dynamic phase portrait.

in a biochemical reaction network. For example, if a system is dynamically stable, the dynamic trajectories will converge to a single point in the plane, known as an attracting fixed point. A stable steady-state point would represent a homeostatic state. Conversely, if the system is unstable, the trajectories will not approach a fixed point but diverge away from it. The former is essentially always the case for biochemical reaction networks representing real cells. The way the trajectories converge on the steady state is highly informative, as different dynamic characteristics are evident from the trajectory.

Characteristic features of phase portraits   A trajectory in the phase portrait may indicate the presence of one or more general dynamic features. Namely, the shapes of the trajectories contain significant information about the dynamic characteristics of a network. Some important features of trajectories in a phase portrait are shown in Figure 3.4.

1. When the trajectory has a negative slope, it indicates that one concentration is increasing while the other is decreasing. The concentrations are moving on the same time scales but in opposite directions; that is, one is consumed while the other is produced. This feature might represent the substrate concentration versus the product concentration of a given reaction. Such behavior helps define aggregate concentration variables.

2. When a trajectory in the phase portrait between two concentrations is a straight line with a positive slope, it means that the two concentrations are moving in tandem; that is, as one increases, so does the other. This feature is observed when two or more concentrations move on the same time scales and are in quasi-equilibrium with one another. Such behavior helps define aggregate concentration variables.

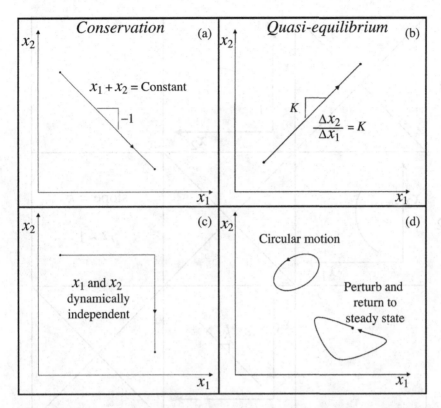

Figure 3.4 General features of dynamic phase portraits. Dynamic phase portraits are formed by graphing the time-dependent concentrations of two concentrations (x_1 and x_2) against one another. Phase portraits have certain characteristic features. (a) Conservation relationship. (b) A pair of concentrations that could be in quasi-equilibrium with one another. (c) Motion of the two concentrations dynamically independent of one another. (d) Closed-loop traces representing either a periodic motion or a return to the original steady state. Modified from [64].

3. When a trajectory is vertical or horizontal, it indicates that one of the concentrations is changing while the other remains constant. This feature implies either that the motions of the concentrations during the trajectory are independent of one another or that the dynamic motions of the concentrations progress on different characteristic time scales. Such behavior helps define time-scale decomposition.

4. When a trajectory forms a closed loop, it implies one of two possibilities. The system never converges to a steady state over time but oscillates, forming a closed-loop trajectory. On the other hand, if the orbit begins at one point, moves away from it, and then returns to the same point after a sufficiently long time interval, then it implies that a change in another variable in the system forced it away from its steady state temporarily, but returned to the original steady state. Such behavior helps define disturbance rejection characteristics.

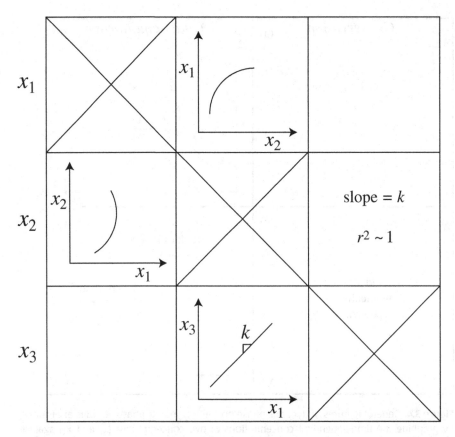

Figure 3.5 A schematic of a tiled phase portrait. The matrix is symmetric, making it possible to display statistical information about a phase portrait in the mirror position. The diagonal elements are meaningless. Originally developed in [64].

The qualitative characteristics of dynamic phase portraits can provide insight into the dynamic features of a network. A trajectory may have more than one of these basic features. For instance, there can be a fast independent motion (i.e., a horizontal phase portrait trajectory) followed by a line with a positive slope after an equilibrium state has been reached.

Tiling dynamic phase portraits   Phase portraits show the dynamic relationships between two variables on multiple time scales; see Figure 3.5. If a system has n variables, then there are n^2 dynamic phase portraits. All pairwise phase portraits can be tiled in a matrix form where the i, j entry represents the dynamic phase portrait between variables x_i and x_j. Note that such an array is symmetric and that the diagonal elements are uninformative. Thus, there are $(n^2 - n)/2$ phase portraits of interest. This feature of this graphical representation opens the possibility of putting the

phase portrait in the i, j position in the array and showing other information (such as a regression coefficient or a slope) in the corresponding j, i
position.

Since the time scales in biochemical reaction networks typically vary
over many orders of magnitude, it often makes sense to make a series
of tiled phase portraits, each of which represents a key time scale. For
instance, rapid equilibration leads to straight lines with positive slopes
in the phase portrait (Figure 3.4b), where the slope is the equilibrium
constant of the reaction. This may be one of many dynamic events talking
place. If a phase portrait is graphed separately on this time scale alone,
then the positive line will show up with a high regression coefficient and
a slope that corresponds to the equilibrium constant. Figures 3.8 and 3.9
in this chapter demonstrate a simple example of the use of tiled phase
portraits. More complex cases are shown in later chapters (see Figures 10.4
and 10.6), where we show how concentration and flux phase portraits are
tiled on multiple time scales to get an overall view of the dynamics of a
whole pathway.

3.3 Post-processing the solution

The initial suggestions obtained from graphing and visualizing the concentration vector $\mathbf{x}(t)$ can lead to a more formal analysis of the results. We
describe three post-processing procedures of $\mathbf{x}(t)$.

1. *Computing the fluxes from the concentration variables* The solution for
the concentrations $\mathbf{x}(t)$ can be used to compute the fluxes from

$$\mathbf{v}(t) = \mathbf{v}(\mathbf{x}(t)) \qquad (3.3)$$

and subsequently we can plot the fluxes in the same way as the concentrations. Graphical information about both the $\mathbf{x}(t)$ and $\mathbf{v}(t)$ is useful.

2. *Combining concentrations to form aggregate variables* The graphical
and statistical multi-time-scale analysis discussed above may lead to the
identification of aggregate variables. Pooled variables $\mathbf{p}$ are computed from

$$\mathbf{p}(t) = \mathbf{P}\mathbf{x}(t), \qquad (3.4)$$

where the pool transformation matrix $\mathbf{P}$ defines the linear combination of
the concentration variables that forms the aggregate variables. For instance,
if we find that a logical way to pool two variables, x_1 and x_2, into new
aggregate variables is $p_1 = x_1 + x_2$ and $p_2 = x_1 - x_2$, then we form the

following matrix equation describing these relationships:

$$\mathbf{p}(t) = \mathbf{P}\mathbf{x}(t) = \begin{pmatrix} p_1(t) \\ p_2(t) \end{pmatrix} = \begin{pmatrix} 1 & 1 \\ 1 & -1 \end{pmatrix} \begin{pmatrix} x_1(t) \\ x_2(t) \end{pmatrix} = \begin{pmatrix} x_1(t) + x_2(t) \\ x_1(t) - x_2(t) \end{pmatrix}.$$

The dynamic variables $\mathbf{p}(t)$ can be studied graphically as described in Section 3.2.

Example: the phosphorylated adenosines. The pool formation discussed in Chapter 2 can be described by the following pool transformation matrix:

$$\mathbf{P} = \begin{pmatrix} 1 & 1 & 0 \\ 2 & 1 & 0 \\ 1 & 1 & 1 \end{pmatrix}, \tag{3.5}$$

and thus

$$\mathbf{p} = \mathbf{P}\mathbf{x} = \mathbf{P} \begin{pmatrix} \text{ATP} \\ \text{ADP} \\ \text{AMP} \end{pmatrix} = \begin{pmatrix} \text{ATP} + \text{ADP} \\ 2\,\text{ATP} + \text{ADP} \\ \text{ATP} + \text{ADP} + \text{AMP} \end{pmatrix}. \tag{3.6}$$

The pool sizes $p_i(t)$ can then be graphed as a function of time.

3. *Correlating concentrations over time* One can construct the time-separated correlation matrix $\mathbf{R}$ based on a time-scale structure of a system. In this matrix, we compute the correlation between two concentrations on a time scale as

$$\mathbf{R}(\tau) = (r_{ij}) = \frac{\sigma_{ij}(\tau)}{\sqrt{\sigma_{ii}\sigma_{jj}}}, \tag{3.7}$$

in which σ_{ii} is the variance of the dataset $x_i(k)$ and $\sigma_{ij}(\tau)$ is the time-lagged covariance between the discrete, uniformly sampled datasets $x_i(k)$ and $x_j(k+\tau)$, determined as

$$\sigma_{ij}(\tau) = \frac{1}{n} \sum_{k=1}^{n-\tau} (x_i(k) - \bar{x}_i)(x_j(k+\tau) - \bar{x}_j), \tag{3.8}$$

in which n is the number of data points in the series and $\bar{x}_i$ indicates the average value of the series x_i. The values in $\mathbf{R}$ range from -1 to $+1$, indicating perfect anti-correlation or correlation, respectively, between two datasets with a delay of time steps. Elements in $\mathbf{R}$ equal to zero indicate that the two corresponding datasets are completely uncorrelated. If such correlation computations were done for the cases shown in Figure 3.4, then one would expect to find a strong negative correlation for the data shown in Figure 3.4a, a strong positive correlation for Figure 3.4b, and no correlation for Figure 3.4c.

The correlation computations can be performed with an increment τ offset in time between two concentrations. An example of a time offset is shown in Figure 3.1 showing the values used from the output file to compute the correlation between x_1 and x_2 with a time lag of 2.

The matrix of phase portraits is symmetric with uninformative diagonal elements. One can, therefore, enter a correlation coefficient corresponding to a particular phase portrait in the transpose position to the phase portrait in the matrix; see Figure 3.8. A correlation coefficient provides a quantitative description of the phase portrait's linearity between the two variables over the time scale displayed. In addition to the correlation coefficient, the slope can be computed and displayed, giving the equilibrium constant between the two compounds displayed.

3.4 Demonstration of the simulation procedure

The simulation procedure and the utility of these graphical approaches will now be illustrated though the use of an example that will hopefully be fairly easy to conceptualize. We will examine a series of linear reversible reactions that ends with the irreversible removal of the last product. The four steps described in Section 3.1 are now outlined.

Step 1: Formulating the dynamic mass balances The reactions considered are

$$x_1 \overset{v_1}{\rightleftharpoons} x_2 \overset{v_2}{\rightleftharpoons} x_3 \overset{v_3}{\rightarrow} x_4. \tag{3.9}$$

The stoichiometric matrix is

$$S = \begin{pmatrix} -1 & 0 & 0 \\ 1 & -1 & 0 \\ 0 & 1 & -1 \\ 0 & 0 & 1 \end{pmatrix}. \tag{3.10}$$

There are four variables and three reactions. Only three of the variables are dynamically independent. One can readily see that the sum of the rows in S is zero. Thus there is a one-dimensional left null space that is spanned by $(1,1,1,1)$ that corresponds to $x_1 + x_2 + x_3 + x_4$ being a constant. Furthermore, x_4 does not influence the dynamics of the reaction chain since v_3 is irreversible.

Step 1: define the equations

```
x1'[t] = b1 - v1; x2'[t] = v1 - v2; x3'[t] = v2 - v3; x4'[t] = v3;
```

Step 2: define the parameter values and rate laws

```
k1 = 1; k2 = 0.01; k3 = 0.0001;
K1 = 1.0; K2 = 1.0;
v1 = k1 * (x1[t] - x2[t] / K1);
v2 = k2 * (x2[t] - x3[t] / K2);
v3 = k3 * x3[t];
b1 = 0;
```

Step 3: set up the conditions and simulate

```
tfinal = 50000;

solution = NDSolve[{x1'[t] = b1 - v1, x2'[t] = v1 - v2, x3'[t] = v2 - v3, x4'[t] = v3,
    x1[0] = 1.0, x2[0] = 0, x3[0] = 0, x4[0] = 0.}, {x1, x2, x3, x4}, {t, 0, tfinal}];
```

Step 4: graph the results

```
<< Graphics`Legend`

graph1 = Plot[{x1[t] /. solution, x2[t] /. solution, x3[t] /. solution, x4[t] /. solution},
    {t, 0.01, 3}, PlotLabel -> "Concentrations vs time",
    PlotLegend -> {"x1", "x2", "x3", "x4"}, AxesLabel -> {"Time", " "},
    PlotStyle -> {Hue[0.5], Hue[0.6], Hue[0.8], Hue[1.0]}, PlotRange -> {0, 1.0},
    LegendPosition -> {1.1, -.4}, LegendSize -> {.35, .35}];

graph2 = Plot[{x1[t] /. solution, x2[t] /. solution, x3[t] /. solution, x4[t] /. solution},
    {t, 3, 300}, PlotLabel -> "Concentrations vs time",
    PlotLegend -> {"x1", "x2", "x3", "x4"}, AxesLabel -> {"Time", " "},
    PlotStyle -> {Hue[0.5], Hue[0.6], Hue[0.8], Hue[1.0]}, PlotRange -> {0, 1.0},
    LegendPosition -> {1.1, -.4}, LegendSize -> {.35, .35}];

graph3 = Plot[{x1[t] /. solution, x2[t] /. solution, x3[t] /. solution,
    x4[t] /. solution},
    {t, 300, tfinal}, PlotLabel -> "Concentrations vs time",
    PlotLegend -> {"x1", "x2", "x3", "x4"}, AxesLabel -> {"Time", " "},
    PlotStyle -> {Hue[0.5], Hue[0.6], Hue[0.8], Hue[1.0]}, PlotRange -> {0, 1.0},
    LegendPosition -> {1.1, -.4}, LegendSize -> {.35, .35}];

graph4 = ParametricPlot[Evaluate[{x1[t], x2[t]} /. solution],
    {t, 0, tfinal}, PlotLabel -> "Dynamic Phase portrait", PlotStyle -> {Hue[0.5]},
    AxesLabel -> {"x1", "x2"}, PlotRange -> {{0, 1}, {0, 1}}];
```

Figure 3.6 A Mathematica workbook that solves the equations for $x_1 \rightleftharpoons x_2 \rightleftharpoons x_3 \rightarrow x_4$ and graphs the solution. The first three graphs are the basis for Figure 3.7.

We can now use $\mathbf{S}$ in Eq. (1.1) to form the dynamic mass balances:

$$\frac{d\mathbf{x}}{dt} = \mathbf{Sv(x)} = \begin{pmatrix} -1 & 0 & 0 \\ 1 & -1 & 0 \\ 0 & 1 & -1 \\ 0 & 0 & 1 \end{pmatrix} \begin{pmatrix} v_1 \\ v_2 \\ v_3 \end{pmatrix} = \begin{pmatrix} -v_1 \\ v_1 - v_2 \\ v_2 - v_3 \\ v_3 \end{pmatrix}. \qquad (3.11)$$

Step 2: Specify the rate laws and parameter values The reactions are elementary and the corresponding rate laws are

$$v_1 = k_1(x_1 - x_2/K_1), \quad v_2 = k_2(x_2 - x_3/K_2), \quad v_3 = k_3 x_3. \qquad (3.12)$$

The parameter values that we will use in this example are $k_1 = 1.0$, $k_2 = 0.01$, $k_3 = 0.0001$, $K_1 = 1$, and $K_2 = 1$. The first reaction has a response time that is 100 times faster than the first, and the second reaction is 100 times faster than the second. The equilibrium constants of unity will lead to an equal value for the concentrations if the reaction is at equilibrium.

Step 3: Numerical simulation The dynamics of this chain of reactions can be simulated for a given condition. Here, we do so from a starting point where only x_1 is present, with $x_1(0) = 1$ and all other concentrations being zero at $t = 0$. This simulation can be accomplished using a variety of software packages that are readily available. A Mathematica workbook that solves these equations is shown in Figure 3.6.

Step 4: Graphical representation The results of the numerical integration can be graphed and post-processed. We can: (1) present the time profiles; (2) tile the phase portraits; and (3) form pools, or aggregate variables, over time.

(1) *Time profiles.* The time profiles can be shown in different ways, recall Figure 3.2. In this case we have chosen to show the time profiles of all the concentrations simultaneously and to partition the x-axis into three separate time regimes scaled by the rate constants of the three reactions. The results are shown in Figure 3.7. These results can be inspected and interpreted. The dynamics that unfold on these three time scales are as follows:

1. On the fastest time scale, we see that x_1 equilibrates with x_2 without much conversion of x_2 to x_3. The first time region of Figure 3.7 shows that x_1 goes from the initial value of 1 to about 0.5 and x_2 rises from

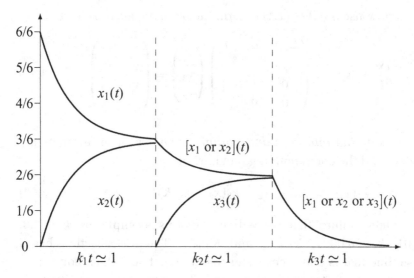

Figure 3.7 Dynamic response of a linear sequence of reactions; $x_1 \rightleftharpoons x_2 \rightleftharpoons x_3 \rightarrow x_4$. The parameter values are $k_1 = 1.0$, $k_2 = 0.01$, $k_3 = 0.0001$, $K_1 = 1$, $K_2 = 1$, $x_1(0) = 1$, and $x_2(0) = x_3(0) = x_4(0) = 0$.

0 to about 0.5. The concentrations of x_3 and x_4 remain effectively zero. Thus, x_1 and x_2 reach a quasi-equilibrium state in which they effectively form a dynamic aggregate of $(x_1 + x_2)$.

2. On the intermediate time scale, we see that the dynamic aggregate of $(x_1 + x_2)$ equilibrates with x_3. The aggregate $(x_1 + x_2)$ moves from a value of about 1.0 to about $\frac{2}{3}$, while the concentration of x_3 rises from 0 to about $\frac{1}{3}$ (Figure 3.7). Thus, on this time scale we see v_2 reach a quasi-equilibrium state, at the end of which the ratio between x_1, x_2, x_3 is about 1:1:1, or a concentration of about $\frac{1}{3}$ each (the total remains 1.0). An aggregate variable of $x_1 + x_2 + x_3$ forms on this time scale.

3. On the slowest time scale, the aggregate variable of $x_1 + x_2 + x_3$ decays to form x_4 (Figure 3.7). All the concentrations move in tandem on this time scale and decay from $\frac{1}{3}$ to zero, while x_4 builds up to a concentration of unity.

Thus, the dynamic response unfolds on multiple time scales and the events on each time scale can be interpreted in chemical terms and analyzed on a time-separated series of tiled phase portraits.

(2) *Phase portraits.* This sequence of events can be further examined using dynamic phase portraits. Since dynamic relationships form on particular time scales, to further study the dynamic response we plot all pairwise tiled phase portraits; see Figure 3.8.

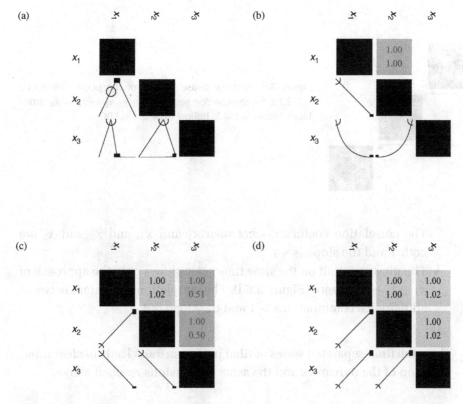

Figure 3.8 A tiled set of dynamic phase portraits for $x_1 \rightleftharpoons x_2 \rightleftharpoons x_3 \to x_4$. The dynamic response is described in the text. (a) The entire response ($t = 0$ to $400\,000$); small black square is $t = 3$; hollow circle is $t = 300$. (b) The fast time scale ($t = 0$ to 1); (c) the middle time scale ($t = 3$ to 100); (d) the slow time scale ($t = 300$ to $10\,000$). Note that, in (b), the scale for x_3 is expanded to show the small change that occurs in its concentration.

- The overall view of the transient response shows the three time regimes; see Figure 3.8a. The pair of x_1 and x_2 moves rapidly along a $-45°$ line to the quasi-equilibrium line on the fast time scale, that is the $45°$ line ($K_1 = 1$) then on the intermediate and slow time scales they move in a fixed ratio of $x_1/x_2 = 1$. x_3 does not move much on the fast time scale, but on the intermediate time scale it equilibrates with x_1 and x_2, followed by a motion along the $45°$ line on the slow time scale relative to both x_1 and x_2.
- The phase portrait on the fast time scale shows only the approach of x_1 and x_2 to quasi-equilibrium (Figure 3.8b). The correlation coefficient is 1 and the slope is computed to be -1. The motion of x_3 is insignificant in quantitative terms.
- The phase portrait on the intermediate time scale shows only the approach of x_3 to quasi-equilibrium with $x_1 + x_2$ (Figure 3.8c). The correlation coefficient between x_1 and x_2 is 1 and the slope is 1.

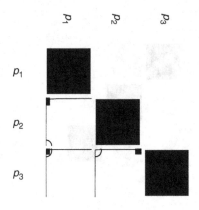

Figure 3.9 Dynamic phase portrait of the pools defined by Eq. (3.13) for the reaction sequence $x_1 \rightleftharpoons x_2 \rightleftharpoons x_3 \rightarrow x_4$; small black square is $t = 3$; hollow circle is $t = 300$.

The correlation coefficients between x_1 and x_3, and x_2 and x_3 are both 1 and the slope is -1.

- The phase portrait on the slow time scale shows only the approach of $x_1 + x_2 + x_3$ to zero (Figure 3.8d). The correlation coefficient between all pairwise combinations is 1 and the slope is 1.

Thus, the time-separated series of tiled phase portraits leads to clear interpretation of the dynamics and the same conclusions reached above.

(3) *Forming pools.* Based on the interpretation of the multiscale dynamic response outlined above, we postulate that the following pooling matrix will temporally decompose the dynamic response:

$$\mathbf{P} = \begin{pmatrix} 1 & -1 & 0 \\ 0.5 & 0.5 & -1 \\ 1 & 1 & 1 \end{pmatrix}, \tag{3.13}$$

where the first two pools describe the equilibration of the first two reactions, $p_1 = x_1 - x_2$ and $p_2 = (x_1 + x_2)/2 - x_3$, and the third describes the motion of the pool of $p_3 = x_1 + x_2 + x_3$. Once these pools are graphed as dynamic variables on dynamic phase portraits they show L-shaped curves (Figure 3.9), demonstrating their dynamic decoupling; see Figure 3.4c.

3.5 Summary

➤ Network dynamics are described by dynamic mass balances ($d\mathbf{x}/dt = \mathbf{Sv}(\mathbf{x}; \mathbf{k})$) that are formulated after applying a series of simplifying assumptions (e.g., Table 1.2).

➤ To simulate the dynamic mass balances we have to specify the numerical values of the kinetic constants $\mathbf{k}$, the initial conditions $\mathbf{x}_0$, and any fixed boundary fluxes.

➤ The equations with the initial conditions can be integrated numerically using a variety of available software packages.

➤ The solution contains numerical values for the concentration variables at discrete time points. The solution is graphically displayed as concentrations over time, or in a phase portrait.

➤ The solution can be post-processed following its initial analysis to bring out special dynamic features of the network. We will describe such features in more detail in the following three chapters.

## CHAPTER 4

Chemical reactions

The simulation procedure described in the previous chapter is now applied to a series of simple examples that represent chemical reactions. We first remind the reader of some key properties of chemical reactions that will show up in dynamic simulations and determine characteristics of network dynamic responses. We then go through a set of examples of chemical reactions that occur in a *closed system*. A closed system is isolated from its environment. No molecules enter or leave the system. Reactions being carried out in the laboratory in a sealed container represent an example of closed systems. In this chapter we assign numerical values to all the parameters for illustration purposes. In Chapter 7 we discuss the numerical values of the various quantities that appear in simulation models of biochemical reaction networks.

4.1 Basic properties of reactions

Links between molecular components in a biochemical reaction network are given by chemical reactions or associations between chemical components. These links, therefore, are characterized and constrained by basic chemical rules.

Bilinear reactions are prevalent in biology Although there are linear reactions found in biological reaction networks, the prototypical transformations in living systems at the molecular level are bilinear. This association involves two compounds coming together to either be chemically transformed through the breakage and formation of a new covalent bond, as is typical of metabolic reactions or macromolecular synthesis,

$$X + Y \rightleftharpoons X - Y \quad \text{covalent bonds,}$$

or two molecules associated together to form a complex that may be held together by hydrogen bonds and/or other physical association forces to form a complex that has a different functionality than individual components,

$$X + Y \rightleftharpoons X : Y \quad \text{association of molecules.}$$

Such association, for instance, could designate the binding of a transcription factor to DNA to form an activated site to which an activated polymerase binds. Such bi-linear association between two molecules might also involve the binding of an allosteric regulator to an allosteric enzyme that induces a conformational change in the enzyme.

Properties of biochemical reactions   Chemical transformations have three key properties that will influence the dynamic features of reaction networks and how we interpret dynamic states:

1. *Stoichiometry.* The stoichiometry of chemical reactions is fixed and is described by integral numbers counting the molecules that react and that form as a consequence of the chemical reaction. Thus, stoichiometry basically represents "digital information." Chemical transformations are constrained by elemental and charge balancing, as well as other features. Stoichiometry is invariant between organisms for the same reactions and does not change with pressure, temperature, or other conditions. Stoichiometry gives the primary topological properties of a biochemical reaction network.

2. *Thermodynamics.* All reactions inside a cell are governed by thermodynamics that determine the equilibrium state of a reaction. The relative rates of the forward and reverse reactions, therefore, are fixed by basic thermodynamic properties. Unlike stoichiometry, thermodynamic properties do change with physico-chemical conditions such as pressure and temperature. Thus, the thermodynamics of transformation between small molecules in cells are fixed but condition dependent. The thermodynamic properties of associations between macromolecules can be changed by altering the amino acid sequence of a protein or by phosphorylation of amino acids in the interface region, or by conformational change induced by the binding of a small-molecule ligand.

3. *Absolute rates.* In contrast to stoichiometry and thermodynamics, the absolute rates of chemical reactions inside cells are highly manipulable. Highly evolved enzymes are very specific in catalyzing particular chemical transformations (recall Figure 2.4). Cells can thus extensively manipulate the absolute rates of reactions through changes in their DNA sequence.

All biochemical transformations are subject to the basic rules of chemistry and thermodynamics.

4.2 The reversible linear reaction

We start with the reversible linear reaction:

$$x_1 \underset{v_{-1}}{\overset{v_1}{\rightleftharpoons}} x_2. \tag{4.1}$$

Here, we have that

$$S = \begin{pmatrix} -1 & 1 \\ 1 & -1 \end{pmatrix}, \quad v(x) = \begin{pmatrix} v_1(x_1) \\ v_{-1}(x_2) \end{pmatrix} = \begin{pmatrix} k_1 x_1 \\ k_{-1} x_2 \end{pmatrix},$$

and thus the differential equations that we will need to simulate are

$$\frac{dx_1}{dt} = -k_1 x_1 + k_{-1} x_2, \quad \frac{dx_2}{dt} = k_1 x_1 - k_{-1} x_2 = -\frac{dx_1}{dt}, \tag{4.2}$$

with the reaction rate given as the difference between two elementary reaction rates

$$v_{1,net} = v_1 - v_{-1} = k_1 x_1 - k_{-1} x_2 = k_1(x_1 - x_2/K_1), \tag{4.3}$$

where $K_1 = k_1/k_{-1}$ ($= x_{2,eq}/x_{1,eq}$) or the ratio of the product to reactant concentrations at equilibrium, the conventional definition of an equilibrium constant in chemistry. Note that, in Eq. (4.3), k_1 represents the kinetics, or the rate of change, while $(x_1 - x_2/K_1)$ represents the thermodynamics measuring how far from equilibrium the system is, i.e., $x_{1,eq} - x_{2,eq}/K_1 = 0$; see below for details.

A simulation of the dynamic mass balances can be implemented in Mathematica or a similar software package. A sample solution is shown in Figure 4.1, for $k_1 = 1$ and $k_{-1} = 2$. These simulation results can be examined further, and they reveal three important observations: (1) the existence of a conservation quantity; (2) a thermodynamic driving force; and (3) the pooling of variables based on chemistry and thermodynamics.

(1) *Mass conservation.* The time profiles in Figure 4.1b show x_1 fall and x_2 rise to their equilibrium values. The phase portrait (Figure 4.1a) is a straight line of slope -1, of the sort that is shown in Figure 3.4a. This implies that

$$p_2 = x_1 + x_2 = \langle (1, 1), (x_1, x_2)^T \rangle \tag{4.4}$$

is a constant. This summation represents a conservation quantity that stems from the fact that, as x_1 reacts, x_2 appears in an equal and opposite amount. The stoichiometric matrix is singular with a rank of 1, showing

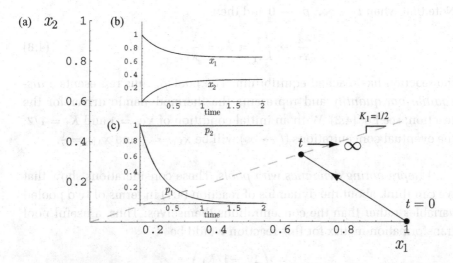

Figure 4.1 Dynamic simulation of the reaction $x_1 \rightleftharpoons x_2$ for $k_1 = 1$, $k_{-1} = 2$ and $x_1(0) = 1$, $x_2(0) = 0$. (a) The phase portrait. (b) The time profiles. (c) The time profile of the pooled variables $p_1 = x_1 - x_2/K_1$ and $p_2 = x_1 + x_2$.

that this is a one-dimensional dynamic system. It has a left null space that is spanned by the vector $(1, 1)$, i.e., $(1, 1) \cdot \mathbf{S} = 0$, thus p_2 is in the left null space of $\mathbf{S}$.

We also note that, since $x_1 + x_2$ is a constant, we can describe the concentration of x_1 as a fraction of the total mass, i.e.,

$$f_1 = \frac{x_1}{x_1 + x_2} = \frac{x_1}{p_2}.$$

Pool sizes and the fraction of molecules in a particular state will be used later in the text to define physiologically useful quantities.

(2) *Disequilibrium and the thermodynamic driving force.* A pooled variable

$$p_1 = x_1 - x_2/K_1 \tag{4.5}$$

can be formed (see Figure 4.1c). Combination of the differential equations for x_1 and x_2 leads to

$$\frac{dp_1}{dt} = -(k_1 + k_{-1})p_1, \tag{4.6}$$

and thus the time constant for this reaction is

$$\tau_1 = \frac{1}{k_1 + k_{-1}}. \tag{4.7}$$

Note that when $t \to \infty$, $p_1 \to 0$ and then

$$\frac{x_2}{x_1} \to \frac{k_1}{k_{-1}} = K_1 = \frac{x_{2,eq}}{x_{1,eq}}, \tag{4.8}$$

the reaction has reached equilibrium. The pool p_1 thus represents a *disequilibrium quantity* and represents the thermodynamic driver for the reaction; see Eq. (4.3). With an initial condition of $x_{1,0} = 1$ and $K_1 = 1/2$, the eventual concentrations ($t \to \infty$) will be $x_{1,eq} = \frac{2}{3}$ and $x_{2,eq} = \frac{1}{3}$.

(3) *Representing dynamics with pools.* These considerations show that we can think about the dynamics of reaction (4.1) in terms of two pooled variables rather than the concentrations themselves. Thus, a useful pool transformation matrix for this reaction would be

$$\mathbf{P} = \begin{pmatrix} 1 & -1/K_1 \\ 1 & 1 \end{pmatrix}, \tag{4.9}$$

leading to disequilibrium (p_1) and conservation (p_2) quantities associated with the reaction in Eq. (4.1). The former quantity moves on the time scale given by τ_1 while the latter is time invariant. For practical purposes, the dynamics of the reaction have relaxed within a time duration of three to five times τ_1 (see Figure 4.1b).

The differential equations for the pools can be obtained as

$$\mathbf{P}\frac{d\mathbf{x}}{dt} = \frac{d}{dt}\begin{pmatrix} p_1 \\ p_2 \end{pmatrix} = -(k_1 + k_{-1})\begin{pmatrix} x_1 - x_2/K_1 \\ 0 \end{pmatrix} = -(k_1 + k_{-1})\begin{pmatrix} p_1 \\ 0 \end{pmatrix}.$$

Therefore, the conservation quantity is a constant (time derivative is zero) and the disequilibrium pool is driven by a thermodynamic driving force that is itself multiplied by $-(k_1 + k_{-1})$, that is the inverse of the time constant for the reaction. Thus, the three key features of chemical reactions (Section 4.1), the stoichiometry, thermodynamics, and kinetics, are separately accounted for.

4.3   The reversible bilinear reaction

The reaction mechanism for the reversible bilinear reaction is

$$x_1 + x_2 \underset{v_{-1}}{\overset{v_1}{\rightleftharpoons}} x_3, \tag{4.10}$$

where the elementary reaction rates are

$$v_1 = k_1 x_1 x_2, \quad \text{and} \quad v_{-1} = k_{-1} x_3. \tag{4.11}$$

The forward rate v_1 is a nonlinear function, or more specifically a bilinear function. The variable

$$p_1 = x_1 x_2 - x_3/K_1 \qquad (4.12)$$

represents a disequilibrium quantity. The dynamic states of this system can be computed from

$$\frac{dx_1}{dt} = v_1 - v_{-1} = k_1 x_1 x_2 - k_{-1} x_3 = k_1(x_1 x_2 - x_3/K_1) = \frac{dx_2}{dt} = -\frac{dx_3}{dt}.$$

$$(4.13)$$

This example will be used to illustrate the essential features of a bilinear reaction: (1) that there are two conservation quantities associated with it; (2) how to compute the equilibrium state; (3) the use of linearization and deviation variables from the equilibrium state; (4) the derivation of a single linear disequilibrium quantity; and (5) formation of pools.

(1) *Conservation quantities.* The stoichiometric matrix is

$$S = \begin{pmatrix} -1 & 1 \\ -1 & 1 \\ 1 & -1 \end{pmatrix}.$$

The stoichiometric matrix has a rank of 1, and thus the dynamic dimension of this system is 1. Two vectors that span the left null space of S are $(1, 0, 1)$ and $(0, 1, 1)$ and the corresponding conservation quantities are

$$p_2 = x_1 + x_3, \quad p_3 = x_2 + x_3. \qquad (4.14)$$

This selection of conservation quantities is not unique, as one can find other sets of two vectors that span the left null space.

(2) *The equilibrium state.* We can examine the equilibrium state for the specific parameter values to be used for numerical simulation below; see Figure 4.2. At equilibrium, $p_1 \to 0$ and we have that ($K_1 = 1$)

$$x_{1,eq} x_{2,eq} = x_{3,eq} \qquad (4.15)$$

and that

$$x_1(0) = 3 = x_{1,eq} + x_{3,eq}, \quad x_2(0) = 2 = x_{2,eq} + x_{3,eq}. \qquad (4.16)$$

These three equations can be combined to give a second-order algebraic equation,

$$x_{3,eq}^2 - 6x_{3,eq} + 6 = 0, \qquad (4.17)$$

that has a positive root that yields

$$x_{1,eq} = 1.73, \quad x_{2,eq} = 0.73, \quad x_{3,eq} = 1.27. \qquad (4.18)$$

(3) *Linearization and formation of deviation variables.* Equation (4.13) can be linearized around the equilibrium point $\mathbf{x}_{eq} = (1.73, 0.73, 1.27)$ to give

$$\frac{dx_1}{dt} = x_1 x_2 - x_3 \qquad (4.19)$$

$$\rightarrow 0.73(x_1 - 1.73) + 1.73(x_2 - 0.73) - (x_3 - 1.27), \qquad (4.20)$$

where a numerical value of k_1 used is unity.

(4) *The disequilibrium quantity.* Equation (4.20) can be written in terms of deviation variables from the equilibrium state, i.e.,

$$x_i' = x_i - x_{i,eq}, \qquad (4.21)$$

as

$$\frac{dx_1'}{dt} \simeq 0.73x_1' + 1.73x_2' - x_3' = p_1', \qquad (4.22)$$

which simply is the linearized version of the disequilibrium quantity in Eq. (4.12), and we have that

$$\frac{dx_2'}{dt} = \frac{dx_1'}{dt} \quad \text{and} \quad \frac{dx_3'}{dt} = -\frac{dx_1'}{dt}. \qquad (4.23)$$

(5) *Representing dynamics with pools.* We can, therefore, form a pool transformation matrix as

$$\mathbf{P} = \begin{pmatrix} 0.73 & 1.73 & -1 \\ 1 & 0 & 1 \\ 0 & 1 & 1 \end{pmatrix}, \qquad (4.24)$$

where the first pool represents the disequilibrium quantity and the second and third are conservation quantities. Now we transform the deviation variables with this matrix, i.e., $\mathbf{p}' = \mathbf{P}\mathbf{x}'$, and can look at the time derivatives of the pools

$$\frac{d\mathbf{p}'}{dt} = \mathbf{P}\frac{d\mathbf{x}'}{dt} \qquad (4.25)$$

$$= \begin{pmatrix} -3.46 & 3.46 \\ 0 & 0 \\ 0 & 0 \end{pmatrix} (0.73x_1' + 1.73x_2' - x_3') \qquad (4.26)$$

$$= 3.46 \begin{pmatrix} 1 \\ 0 \\ 0 \end{pmatrix} p_1'. \qquad (4.27)$$

This result is similar to that obtained above for the linear reversible reaction. There are two conservation pools and a disequilibrium pool that is

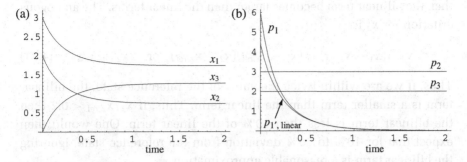

Figure 4.2 The concentration–time profiles for the reaction $x_1 + x_2 \rightleftharpoons x_3$ for $k_1 = k_{-1} = 1$ and $x_1(0) = 3$, $x_2(0) = 2$. (a) The concentrations as a function of time. (b) The pools as a function of time.

moved by itself multiplied by a characteristic rate constant. We note that the conservation quantities, for both the linear and bilinear reactions, do not change if the reactions are irreversible (i.e., if $K_{eq} \to \infty$).

Numerical simulation   The dynamic response of this reaction can readily be computed and the results graphed; see Figure 4.2. Note, in Figure 4.2b, p_1 and p_1' are effectively the same, especially close to equilibrium.

Consequences of linearization   Nonlinear functions are linearized using Taylor series expansion. For functions that are bilinear or quadratic, the expansion is exact after three terms:

$$f(\mathbf{x}) = f(\mathbf{x}_{ref}) + \frac{\partial f}{\partial \mathbf{x}}|_{ref}(\mathbf{x} - \mathbf{x}_{ref}) + \frac{1}{2}(\mathbf{x} - \mathbf{x}_{ref})^T \mathbf{H}(\mathbf{x} - \mathbf{x}_{ref}), \qquad (4.28)$$

where $\mathbf{H}$ is the Hessian matrix. This matrix contains all the second-order derivatives of the function

$$h_{ik} = \frac{\partial^2 f}{\partial x_i \partial x_k} \qquad (4.29)$$

and is symmetric for smooth functions. Bilinear terms can be expanded as

$$x_1 x_2 = x_{1,ref} x_{2,ref} + x_{2,ref}(x_1 - x_{1,ref}) + x_{1,ref}(x_2 - x_{2,ref})$$
$$+ (x_1 - x_{1,ref})(x_2 - x_{2,ref}).$$

The error associated with ignoring the last term and considering only the constant and the linear terms can be estimated by looking at the relative magnitudes of the terms in this equation. If

$$x_{1,ref}(x_2 - x_{2,ref}) < (x_1 - x_{1,ref})(x_2 - x_{2,ref}) \quad \text{or} \quad 2x_{1,ref} < x_1, \qquad (4.30)$$

then the bilinear term becomes larger than the linear terms. The analogous criterion for x_2 is

$$x_{2,\text{ref}}(x_1 - x_{1,\text{ref}}) < (x_1 - x_{1,\text{ref}})(x_2 - x_{2,\text{ref}}) \quad \text{or} \quad 2x_{2,\text{ref}} < x_2. \quad (4.31)$$

Thus, if we are within twice the value of the reference state, the bilinear term is a smaller term than the linear term. Thus, if $x_2/x_{2,\text{ref}} < 0.2$ then the bilinear term is less than 10% of the linear term. One would then expect that for 10% to 20% deviation from the reference state, ignoring the bilinear term is a reasonable approximation.

4.4   Connected reversible linear reactions

Now we consider more than one reaction working simultaneously. We will consider two reversible first-order reactions that are connected by an irreversible reaction:

$$X_1 \underset{V_{-1}}{\overset{V_1}{\rightleftharpoons}} X_2 \overset{V_2}{\rightarrow} X_3 \underset{V_{-3}}{\overset{V_3}{\rightleftharpoons}} X_4. \quad (4.32)$$

The stoichiometric matrix and the reaction vector, are

$$S = \begin{pmatrix} -1 & 1 & 0 & 0 & 0 \\ 1 & -1 & -1 & 0 & 0 \\ 0 & 0 & 1 & -1 & 1 \\ 0 & 0 & 0 & 1 & -1 \end{pmatrix}, \quad v(x) = \begin{pmatrix} k_1 x_1 \\ k_{-1} x_2 \\ k_2 x_2 \\ k_3 x_3 \\ k_{-3} x_4 \end{pmatrix}, \quad (4.33)$$

and thus the dynamic mass balances are

$$\frac{dx_1}{dt} = -k_1 x_1 + k_{-1} x_2,$$

$$\frac{dx_2}{dt} = k_1 x_1 - k_{-1} x_2 - k_2 x_2,$$

$$\frac{dx_3}{dt} = k_2 x_2 - k_3 x_3 + k_{-3} x_4,$$

$$\frac{dx_4}{dt} = k_3 x_3 - k_{-3} x_4.$$

$$(4.34)$$

The net reaction rates are

$$v_{1,\text{net}} = k_1 x_1 - k_{-1} x_2 = k_1 (x_1 - x_2/K_1) \quad (4.35)$$

and

$$v_{3,\text{net}} = k_3 x_3 - k_{-3} x_4 = k_3 (x_3 - x_4/K_3), \quad (4.36)$$

where $K_1 = k_1/k_{-1}$ and $K_3 = k_3/k_{-3}$ are the equilibrium constants. This example can be used to illustrate three concepts: (1) dynamic decoupling; (2) stoichiometric decoupling; and (3) formation of multi-reaction pools.

(1) *Dynamic decoupling through separated time scales.* This linear system can be described by

$$\frac{d\mathbf{x}}{dt} = \mathbf{Jx}, \tag{4.37}$$

where the Jacobian matrix for this system is obtained directly from the equations in (4.34):

$$\mathbf{J} = \begin{pmatrix} -k_1 & k_{-1} & 0 & 0 \\ k_1 & -k_{-1} - k_2 & 0 & 0 \\ 0 & k_2 & -k_3 & k_{-3} \\ 0 & 0 & k_3 & -k_{-3} \end{pmatrix}. \tag{4.38}$$

Note that for linear systems, $\mathbf{x} = \mathbf{x}'$. We observe that the second column in $\mathbf{J}$ is a combination of the second and third columns in $\mathbf{S}$:

$$\begin{pmatrix} j_{12} \\ j_{22} \\ j_{32} \\ j_{42} \end{pmatrix} = \begin{pmatrix} 1 \\ -1 \\ 0 \\ 0 \end{pmatrix} k_{-1} + \begin{pmatrix} 0 \\ -1 \\ 1 \\ 0 \end{pmatrix} k_2 = k_{-1}\mathbf{s}_2 + k_2\mathbf{s}_3. \tag{4.39}$$

The kinetic effects of x_2 are thus felt through both reactions 2 and 3 (i.e., the second and third columns in $\mathbf{S}$ that are the corresponding reaction vectors), $\mathbf{s}_2$ and $\mathbf{s}_3$. These two reaction vectors are weighted by the rate constants (reciprocal of the time constants). Therefore, we expect that if k_2 is numerically much smaller than k_{-1}, then the dynamic coupling is "weak." We consider two sets of parameter values.

- First, we simulate this system with all the rate constants being equal; Figure 4.3. All the concentrations are moving on the same time scale. For a series of reactions, the overall dynamics are expected to unfold on a time scale that is the sum of the individual time constants. Here, this sum is three, and the dynamics have relaxed after a time period of three to five times this value.
- Next, we make the second reaction five times slower than the other two; Figure 4.4a. We see that the two faster reactions come to a quasi-equilibrium state relatively quickly (relative to reaction 3), and form two conservation pools that exchange mass slowly. The sum of the

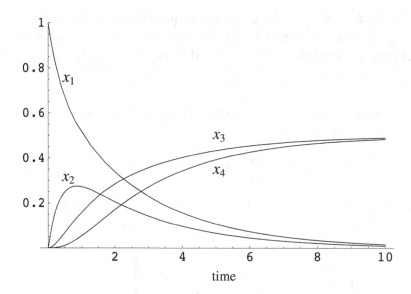

Figure 4.3 The dynamic response of the reactions in Eq. (4.32) for $K_1 = K_3 = 1$ and $k_1 = k_2 = k_3 = 1$. The graphs show the concentrations varying with time for $x_1(0) = 1$, $x_2(0) = x_3(0) = x_4(0) = 0$.

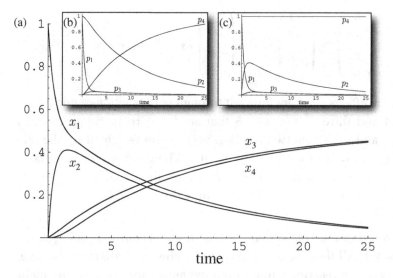

Figure 4.4 (a) The dynamic response of the system in Eq. (4.32) for $K_1 = K_3 = 1$ and $k_1 = 5k_2 = k_3$. The graphs show the concentrations varying with time for $x_1(0) = 1$, $x_2(0) = x_3(0) = x_4(0) = 0$. (b) The conservation pools $p_2 = x_1 + x_2$ and $p_4 = x_3 + x_4$ and the disequilibrium pools $p_1 = x_1 - x_2/K_1$ and $p_3 = x_3 - x_4/K_3$ for the individual reactions. The disequilibrium pools move quickly towards a quasi-equilibrium state, while the conservation pools move more slowly. These pools are defined in Eq. (4.40). (c) The dynamic response with alternative pools; $p_2 = x_2$, and $p_4 = x_1 + x_2 + x_3 + x_4$. These pools are defined in Eq. (4.41).

rate constants for the three reactions in series is now seven, and the dynamics unfold on this time scale.

(2) *Stoichiometric decoupling.* Reaction 3 does not influence reaction 1 at all. They are separated by the irreversible reaction 2. Thus, changes in the kinetics of reaction 3 will not influence the progress of reaction 1. This can be illustrated through simulation by changing the rate constants for reaction 3 and observing what happens to reaction 1.

(3) *Formation of multi-reaction pools.* We can form the following pooled variables based on the properties of the individual reversible reactions (Section 4.2):

$p_1 = x_1 - x_2/K_1$ disequilibrium quantity for reaction 1,
$p_2 = x_1 + x_2$ conservation quantity for reaction 1,
$p_3 = x_3 - x_4/K_3$ disequilibrium quantity for reaction 3,
$p_4 = x_3 + x_4$ conservation quantity for reaction 3.

A representation of the dynamics of this reaction system can be obtained by plotting these pools as a function of time; Figure 4.4b. To prepare this plot, we use the pooling matrix

$$\mathbf{P} = \begin{pmatrix} 1 & -1/K_1 & 0 & 0 \\ 1 & 1 & 0 & 0 \\ 0 & 0 & 1 & -1/K_3 \\ 0 & 0 & 1 & 1 \end{pmatrix} \tag{4.40}$$

to post-process the output (see Section 3.3). However, we note that the conservation quantities associated with the individual reactions are no longer time-invariant.

The rank of $\mathbf{S}$ is 3 and its one-dimensional left null space is spanned by $(1, 1, 1, 1)$; thus, the conservation quantity is $x_1 + x_2 + x_3 + x_4$. Therefore, an alternative pooling matrix may be formulated as

$$\mathbf{P} = \begin{pmatrix} 1 & -1/K_1 & 0 & 0 \\ 0 & 1 & 0 & 0 \\ 0 & 0 & 1 & -1/K_3 \\ 1 & 1 & 1 & 1 \end{pmatrix}, \tag{4.41}$$

where we use x_2 as the coupling variable and the overall conservation pool instead of the conservation pools associated with the individual reactions. The two conservation pools are combined into one overall mass conservation pool.

We can now derive the dynamic mass balances on the pools as

$$\frac{d\mathbf{p}}{dt} = \mathbf{PSv} = \begin{pmatrix} -(k_1 + k_{-1})(x_1 - x_2/K_1) + \frac{k_2}{K_1}x_2 \\ k_1(x_1 - x_2/K_1) - k_2x_2 \\ -(k_3 + k_{-3})(x_3 - x_4/K_3) + k_2x_2 \\ 0 \end{pmatrix}$$

$$= \begin{pmatrix} -(k_1 + k_{-1})p_1 + \frac{k_2}{K_1}p_2 \\ k_1p_1 - k_2p_2 \\ -(k_3 + k_{-3})p_3 + k_2p_2 \\ 0 \end{pmatrix}$$

$$= \begin{pmatrix} -(k_1 + k_{-1}) \\ k_1 \\ 0 \\ 0 \end{pmatrix} p_1 + \begin{pmatrix} \frac{k_2}{K_1} \\ -k_2 \\ k_2 \\ 0 \end{pmatrix} p_2 + \begin{pmatrix} 0 \\ 0 \\ -(k_3 + k_{-3}) \\ 0 \end{pmatrix} p_3.$$

This equation shows that p_1 and p_3 create fast motion compared with p_2 given the relative numerical values of the rate constants; p_2 creates a slow drift in this system for the numerical values used in Figure 4.4c.

4.5   Connected reversible bilinear reactions

An important case of connected bilinear reactions is represented by the reaction mechanism

$$X_1 + X_2 \underset{v_{-1}}{\overset{v_1}{\rightleftharpoons}} X_3 \underset{v_{-2}}{\overset{v_2}{\rightleftharpoons}} X_4 + X_5. \tag{4.42}$$

This reaction network is similar to reaction mechanisms for enzymes, and thus leads us into the treatment of enzyme kinetics (Chapter 5). The elementary reaction rates are

$$v_1 = k_1x_1x_2, \quad v_{-1} = k_{-1}x_3, \quad v_2 = k_2x_3, \quad \text{and} \quad v_{-2} = k_{-2}x_4x_5 \tag{4.43}$$

and the equilibrium constants are $K_1 = k_1/k_{-1}$ and $K_2 = k_2/k_{-2}$. There are two disequilibrium quantities:

$$p_1 = x_1x_2 - x_3/K_1, \tag{4.44}$$

$$p_2 = x_3 - x_4x_5/K_2. \tag{4.45}$$

We now explore the same features of this coupled system of bilinear reactions as we did for the single reversible bilinear reaction; see Section 4.3.

(1) *Conservation quantities.* The (5×4) stoichiometric matrix

$$
S = \begin{pmatrix}
-1 & 1 & 0 & 0 \\
-1 & 1 & 0 & 0 \\
1 & -1 & -1 & 1 \\
0 & 0 & 1 & -1 \\
0 & 0 & 1 & -1
\end{pmatrix}
\tag{4.46}
$$

is of rank 2 and thus there are three conservation variables and two independent dynamic variables.

The conservation quantities are not unique, and which one we will use depends on the reaction chemistry that is being studied. An example is

$$
AB + C \underset{v_{-1}}{\overset{v_1}{\rightleftharpoons}} ABC \underset{v_{-2}}{\overset{v_2}{\rightleftharpoons}} A + BC,
\tag{4.47}
$$

in which case the three independent conservation quantities would be:

$$
\begin{array}{lll}
\text{conservation of A:} & p_3 = x_1 + x_3 + x_4, & \text{(4.48)} \\
\text{conservation of B:} & p_4 = x_1 + x_3 + x_5, & \text{(4.49)} \\
\text{conservation of C:} & p_5 = x_2 + x_3 + x_5. & \text{(4.50)}
\end{array}
$$

These are convex quantities, as all the coefficients are non-negative (the concentrations are $x_i > 0$). The individual bilinear reactions have two each, but once coupled, the number of conservation quantities drops by one.

(2) *The equilibrium state.* The computation of the equilibrium state involves setting the net fluxes to zero and combining those equations with the conservation quantities to get a set of independent equations. For convenience of illustration, we pick $K_1 = K_2 = 1$ and the equilibrium equations become

$$
x_{1,eq} x_{2,eq} = x_{3,eq} = x_{4,eq} x_{5,eq};
\tag{4.51}
$$

and if we pick $p_3 = p_4 = p_5 = 3$, then the solution for the equilibrium state is simple: $x_{1,eq} = x_{2,eq} = x_{3,eq} = x_{4,eq} = x_{5,eq} = 1$. These equations can also be solved for arbitrary parameter values.

(3) *Linearization and deviation variables.* By linearizing the differential equations around the steady state we obtain

$$\frac{dx_1'}{dt} = \frac{dx_2'}{dt} = -(k_1 x_{2,\mathrm{eq}})x_1' - (k_1 x_{1,\mathrm{eq}})x_2' + k_{-1}x_3',$$

$$\frac{dx_3'}{dt} = (k_1 x_{2,\mathrm{eq}})x_1' + (k_1 x_{1,\mathrm{eq}})x_2' - (k_{-1} + k_2)x_3'$$

$$+ (k_{-2} x_{5,\mathrm{eq}})x_4' + (k_{-2} x_{4,\mathrm{eq}})x_5',$$

$$\frac{dx_4'}{dt} = \frac{dx_5'}{dt} = k_2 x_3' - (k_{-2} x_{5,\mathrm{eq}})x_4' - (k_{-2} x_{4,\mathrm{eq}})x_5',$$

where $x_i' = x_i - x_{i,\mathrm{eq}}$ represent the concentration deviation around equilibrium, $i = 1, 2, 3, 4$, and 5.

(4) *The disequilibrium and conservation quantities.* Similar to Section 4.3, we obtain two pools that represent the disequilibrium driving forces of the two reactions

$$p_1 = x_1 x_2 - x_3/K_1 \approx (x_{2,\mathrm{eq}})x_1' + (x_{1,\mathrm{eq}})x_2' - (1/K_1)x_3' = p_1', \quad (4.52)$$

$$p_2 = x_3 - x_4 x_5/K_2 \approx x_3' - (x_{5,\mathrm{eq}}/K_2)x_4' - (x_{4,\mathrm{eq}}/K_2)x_5' = p_2', \quad (4.53)$$

and the three pools that represent conservation quantities do not change:

$$p_3' = x_1' + x_3' + x_4', \tag{4.54}$$

$$p_4' = x_1' + x_3' + x_5', \tag{4.55}$$

$$p_5' = x_2' + x_3' + x_5'. \tag{4.56}$$

We can thus define the pooling matrix as

$$\mathbf{P} = \begin{pmatrix} x_{2,\mathrm{eq}} & x_{1,\mathrm{eq}} & -1/K_1 & 0 & 0 \\ 0 & 0 & 1 & -x_{5,\mathrm{eq}}/K_2 & -x_{4,\mathrm{eq}}/K_2 \\ 1 & 0 & 1 & 1 & 0 \\ 1 & 0 & 1 & 0 & 1 \\ 0 & 1 & 1 & 0 & 1 \end{pmatrix} \tag{4.57}$$

$$= \begin{pmatrix} 1 & 1 & -1 & 0 & 0 \\ 0 & 0 & 1 & -1 & -1 \\ 1 & 0 & 1 & 1 & 0 \\ 1 & 0 & 1 & 0 & 1 \\ 0 & 1 & 1 & 0 & 1 \end{pmatrix} \tag{4.58}$$

for the particular equilibrium constants and concentrations values given above.

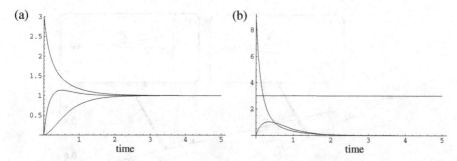

Figure 4.5 The concentration–time profiles for the reaction system $x_1 + x_2 \rightleftharpoons x_3 \rightleftharpoons x_4 + x_5$ for $k_1 = k_{-1} = k_2 = k_{-2} = 1$ and $x_1(0) = 3$, $x_2(0) = 3$, $x_3(0) = 0$, $x_4(0) = 0$, $x_5(0) = 0$. (a) The concentrations as a function of time. (b) The pools as a function of time.

The differential equations for the pools are then formed by (and at this stage we remove the conservation pools as they are always constant)

$$\frac{d\mathbf{p}'}{dt} = \mathbf{PSv(x)}, \quad \text{where} \quad \mathbf{v(x)} \approx \begin{pmatrix} k_1 p_1' \\ k_2 p_2' \end{pmatrix} = \begin{pmatrix} k_1 & 0 \\ 0 & k_2 \end{pmatrix} \begin{pmatrix} p_1' \\ p_2' \end{pmatrix}, \quad (4.59)$$

which gives

$$\frac{d\mathbf{p}'}{dt} = \begin{pmatrix} -(x_{2,eq} + x_{1,eq} + 1/K_1) \\ 1 \end{pmatrix} k_1 p_1' + \begin{pmatrix} 1/K_1 \\ -[1 + (x_{5,eq} + x_{4,eq}]/K_2) \end{pmatrix} k_2 p_2'.$$

(5) *Numerical simulation.* These equations can be simulated once parameter values and initial conditions are specified. In order to illustrate the dynamic behavior in terms of the pools, we consider the particular situation where $K_1 = K_2 = x_{1,eq} = x_{2,eq} = x_{3,eq} = x_{4,eq} = x_{5,eq} = 1$; see Figure 4.5.

In this situation, the dynamic equation for the linearized pools becomes

$$\frac{d\mathbf{p}'}{dt} = \begin{pmatrix} -3 \\ 1 \end{pmatrix} k_1 p_1' + \begin{pmatrix} 1 \\ -3 \end{pmatrix} k_2 p_2'. \quad (4.60)$$

We can solve this equation and present the results with a dynamic phase portrait; Figure 4.6. The dynamic behavior of the nonequilibrium pools is shown for three different sets of parameters. (a) Reaction 1 is 10 times faster than reaction 2, (b) The two reactions have the same response time, and (c) reaction 2 is 10 times faster than reaction 1. We make three observations here.

1. Lines A and C in Figure 4.6 show that the dynamics for the pools can be decomposed to consider a fast equilibration of the two disequilibrium pools followed by the slow decay of the slower disequilibrium pool.

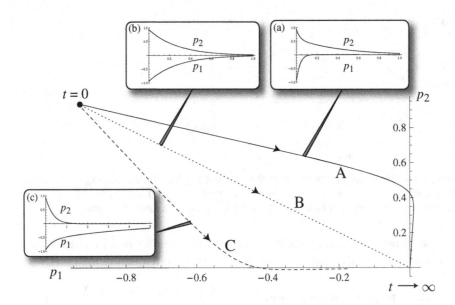

Figure 4.6 The dynamic response of $x_1 + x_2 \rightleftharpoons x_3 \rightleftharpoons x_4 + x_5$ for $K_1 = K_2 = 1$. The graphs show the concentrations varying with time for $x_3(0) = 1.5, x_1(0) = x_2(0) = x_4(0) = x_5(0) = 0.75$. The disequilibrium pools $p_1 = x_1 x_2 - x_3/K_1$ (x-axis) and $p_2 = x_3 - x_4 x_5/K_3$ (y-axis) shown in a phase portrait. (a) The dynamic response of the disequilibrium pools for $k_1 = 10k_2$. (b) The dynamic response of the disequilibrium pools for $k_1 = k_2$. (c) The dynamic response of the disequilibrium pools for $k_1 = 0.1k_2$.

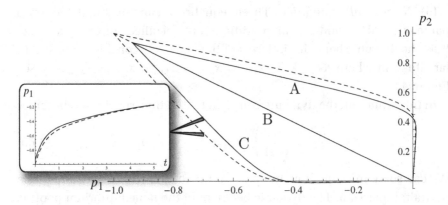

Figure 4.7 The phase portrait for the linearized pools defined in Eqs (4.52) and (4.53). Curve A: $k_1 = 10k_2$. Curve B: $k_1 = k_2$. Curve C: $k_1 = 0.1k_2$. The inset corresponds to p_1 for curve C. Same initial conditions as in Figure 4.6. The dashed lines are the linearized solution and the solid lines are the full solution.

2. When reaction 1 is 10 times faster than reaction 2, then initial motion is along the vector $(-3, 1)^T$, and when reaction 1 is 10 times slower than reaction 2, then initial motion is along the vector $(1, -3)^T$.

3. The linearized pools move in a similar fashion to the bilinear disequilibrium pools; see Figure 4.7. The bilinear and linear simulations

do not change that much even though x_1, x_2, x_4, x_5, are 25% from their equilibrium value, and x_3 is 50% away from equilibrium.

4.6   Summary

➤ Chemical properties associated with chemical reactions are stoichiometry, thermodynamics, and kinetics. The first two are physicochemical properties, while the third can be biologically altered through enzyme action.

➤ Each net reaction can be described by pooled variables that represent a disequilibrium quantity and a mass conservation quantity that is associated with the reaction.

➤ If a reaction is fast compared with its network environment, its disequilibrium variable can be relaxed and then described by the conservation quantity associated with the reaction.

➤ Linearizing bilinear rate laws do not create much error for small changes around the reference state.

➤ Removing a time scale from a model corresponds to reducing the dynamic dimension of the transient response by one.

➤ As the number of reactions grow, the number of conservation quantities may change.

➤ Irreversibility of reactions does not change the number of conservation quantities for a system.

## CHAPTER 5

Enzyme kinetics

We now study common reaction mechanisms that describe enzyme catalysis. Enzymes can dramatically accelerate the rate of biochemical reactions inside and outside living cells. The absolute rates of biochemical reactions are key biological design variables because they can evolve from a very low rate as determined by the mass action kinetics based on collision frequencies, to a very high and specific reaction rate as determined by appropriately evolved enzyme properties. We first describe the procedure used to derive enzymatic rate laws, which we then apply to the Michaelis–Menten reaction mechanism, then to the Hill model, and finally to the symmetry model. The first is used to describe plain chemical transformations, while the latter two are used to describe regulatory effects.

5.1　Enzyme catalysis

Enzymes are catalysts that accelerate biochemical transformations in cells. Almost all enzymes are proteins. There are also catalytically active ribonucleic acids, called "ribozymes." The fundamental properties of enzyme catalysis are described in this section.

Enzymatic activity　The activity of an enzyme is measured by determining the increase in the reaction rate relative to the absence of the enzyme. In other words, we compare the reaction rate of the uncatalyzed reaction with the catalyzed rate. The ratio can be thought of as an acceleration factor and this number can be quite high, sometimes by many million-fold.

Reaction and substrate specificity　Enzymes are usually very specific both with respect to the type of reaction being catalyzed (reaction specificity)

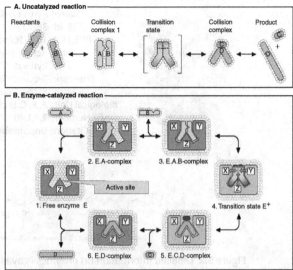

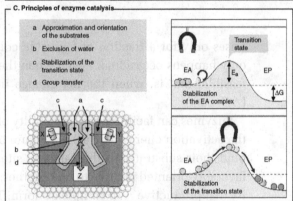

Figure 5.1 Basic principles of enzyme catalysis. From [68] (reprinted with permission).

and with respect to the reactants (the "substrates") that they act on. Highly specific enzymes catalyze the cleavage of only one type of a chemical bond, and only in one substrate. Other enzymes may have a narrow reaction specificity, but broad substrate specificity, i.e., they act on a number of chemically similar substrates. Rare enzymes exist that have both low reaction specificity and low substrate specificity.

Enzyme catalysis As discussed in Chapter 2 (Figure 2.4), two molecules can only react with each other if they collide in a favorable orientation. Such collisions may be rare, and thus the reaction rate is slow. An uncatalyzed reaction starts with a favorable collision, as shown in Figure 5.1A. Before the products are formed, the collision complex A–B has to pass through what is called a *transition state*. Its formation requires *activation energy*. Since activation energies can be quite high, only a few A–B complexes have this amount of energy, and thus a productive transition state

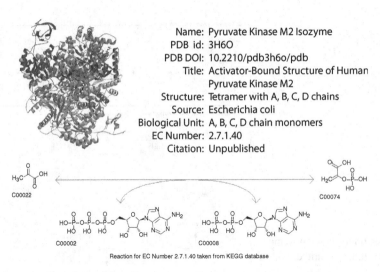

Figure 5.2 Detailed information on enzymes is available. From [1].

arises only for a fraction of favorable collisions. As a result, conversion only happens occasionally even when the reaction is thermodynamically feasible; that is, when the net change in Gibbs free energy is negative ($\Delta G < 0$).

Enzymes can facilitate the probability of a favorable collision and lower the activation energy barrier, see Figure 5.1B and C. Enzymes are able to bind their substrates in the catalytic site. As a result, the substrates are favorably oriented relative to one another, greatly enhancing the probability that productive A–B complexes form. The transition state is stabilized, leading to a lowered activation energy barrier.

Information on enzymes Detailed information is available on a large number of enzymes. This includes structural information, the organism source, and other characteristics. An example is shown in Figure 5.2. Many online sources of such information exist.

5.2   Deriving enzymatic rate laws

The chemical events underlying the catalytic activities of enzymes are described by a *reaction mechanism*. A reaction mechanism is comprised of the underlying elementary reactions that are believed to take place. A rate law is then formulated to describe the rate of reaction.

Step 1: Reaction Mechanics

$$x_1 \rightleftharpoons x_2 \rightarrow x_3$$

$$\frac{d\mathbf{x}}{dt} = \mathbf{S} \cdot \mathbf{v}(\mathbf{x}) = \begin{pmatrix} -1 & 1 & 0 \\ 1 & -1 & -1 \\ 0 & 0 & 1 \end{pmatrix} \begin{pmatrix} v_1 \\ v_2 \\ v_3 \end{pmatrix}$$

Step 2: Time-invariants

$$\mathbf{L} \cdot \mathbf{S} = 0$$
$$x_1 + x_2 + x_3 = a$$

Step 3: Eliminate Dependent Variables

$$x_3 = a - x_1 - x_2$$

Step 4: Simplifying Kinetic Assumptions

$$\frac{dx_1}{dt} = -k_1 x_1 + k_{-1} x_2 \qquad \text{QSSA:} \qquad \frac{dx_2}{dt} = 0 \rightarrow \frac{dx_1}{dt} = -k_2 \left(\frac{k_1}{k_2 + k_{-1}} \right) x_1$$

$$\frac{dx_2}{dt} = k_1 x_1 - (k_{-1} + k_2) x_2 \qquad \text{QEA:} \qquad K_1 = \frac{x_2}{x_1} \rightarrow \frac{dx_2}{dt} = -k_2 x_2$$

Step 5: Simulate Full and Approximate Solution

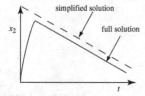

Step 6: Formulate Dimensionless Groups

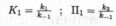

$$K_1 = \frac{k_1}{k_{-1}} \quad ; \quad \Pi_1 = \frac{k_2}{k_{-1}}$$

Figure 5.3 The process of formulating enzymatic rate laws. QSSA represents the quasi-steady state assumption and QEA represents the quasi-equilibrium assumption.

A rate law describes the conversion of a substrate (x_1) by an enzyme into a product (x_2):

$$x_1 \xrightarrow{v} x_2, \tag{5.1}$$

where v is a function of the concentrations of the chemical species involved in the reaction. The steps involved in the development and analysis of enzymatic rate laws are illustrated in Figure 5.3 and they are as follows:

Step 1: formulate the dynamic mass balances based on the elementary reactions in the postulated reaction mechanism.

Step 2: identify time invariants, or conservation relationships.

Step 3: reduce the dynamic dimension of the reaction mechanism by eliminating dynamically dependent variables using the conservation relationships.

Step 4: apply commonly used simplifying kinetic assumptions to formulate a rate law, representing a reduction in the dynamic dimension of the kinetic model.

Step 5: apply mathematical and numerical analysis to determine when the simplifying assumptions are valid and the reaction rate law can be used.

Step 6: identify key dimensionless parameter ratios. This last step is optional and used by those interested in deeper mathematical analysis of the properties of the rate laws.

The use of enzymatic rate laws in dynamic network models is hampered by their applicability *in vivo* based on *in vitro* measurements. From a practical standpoint, with the numerical simulation capacity that is now routinely available, applying simplifying assumptions may no longer be needed for computational simplification and convenience. However, it is useful to help understand the historical origin of enzymatic rate laws, the simplifications on which they are based, and when it may be desirable to use them.

5.3   Michaelis–Menten kinetics

The simplest enzymatic reaction mechanism, first proposed by Henri [43], but named after Michaelis and Menten [80], is

$$S + E \underset{k_{-1}}{\overset{k_1}{\rightleftharpoons}} X \overset{k_2}{\rightarrow} E + P, \tag{5.2}$$

where a substrate S binds reversibly to the enzyme E to form the intermediate X, which can break down to give the product P and regenerate the enzyme. Note that it is similar to the reaction mechanism of two connected reversible bilinear reactions (Eq. (4.42)) with $x_5 = x_2$, as one of the original reactants (E) is regained in the second step. Historically speaking, the Michaelis–Menten scheme is the most important enzymatic reaction mechanism. A detailed account of the early history of Michaelis–Menten kinetics is found in [20].

Step 1: Dynamic mass balances   Applying the law of mass action to the Michaelis–Menten reaction mechanism, one obtains four differential equations that describe the dynamics of the concentrations of the four chemical species involved in the reaction mechanism:

$$\frac{ds}{dt} = -k_1 es + k_{-1} x, \qquad s(t = 0) = s_0,$$

$$\frac{dx}{dt} = k_1 es - (k_{-1} + k_2)x, \qquad x(t = 0) = 0,$$

$$\frac{de}{dt} = -k_1 es + (k_{-1} + k_2)x, \qquad e(t = 0) = e_0,$$

$$\frac{dp}{dt} = k_2 x, \qquad p(t = 0) = 0,$$

(5.3)

where the lower case letters denote the concentrations of the corresponding chemical species. The initial conditions shown are for typical initial rate experiments where substrate and free enzyme are mixed together at time $t = 0$. e_0 and s_0 denote the initial concentration of enzyme and substrate respectively. No mass exchange occurs with the environment.

Step 2: Finding the time invariants Using $\mathbf{x} = (s, e, x, p)$ and $\mathbf{v} = (k_1 es, k_{-1} x, k_2 x)$, the stoichiometric matrix is

$$\mathbf{S} = \begin{pmatrix} -1 & 1 & 0 \\ -1 & 1 & 1 \\ 1 & -1 & -1 \\ 0 & 0 & 1 \end{pmatrix}.$$

(5.4)

It has a rank of 2 and, thus, there are two conservation quantities. They are the total concentration of the enzyme and total concentration of the substrate:

$$e_0 = e + x,$$

(5.5)

$$s_0 = s + x + p.$$

(5.6)

Step 3: Reducing the dynamic description As a consequence of the two conservation relationships, only two of the equations in (5.3) are dynamically independent. Choosing the substrate s and the intermediate complex x concentrations as the two independent variables, the reaction dynamics are described by:

$$\frac{ds}{dt} = -k_1 e_0 s + (k_1 s + k_{-1})x, \quad s(t = 0) = s_0,$$

(5.7)

$$\frac{dx}{dt} = k_1 e_0 s - (k_1 s + k_{-1} + k_2)x, \quad x(t = 0) = 0.$$

(5.8)

The major problem with this mass action kinetic model is that it is mathematically intractable [48]. Equations (5.7) and (5.8) can be reduced to an Abel-type differential equation whose solution cannot be obtained in a closed form.

Step 4: Applying kinetic assumptions   A closed-form analytical solution to the mass action kinetic equations, (5.7) and (5.8), is only attainable by using simplifying kinetic assumptions. Two assumptions are used: the *quasi-steady-state assumption* (QSSA) and the *quasi-equilibrium assumption* (QEA).

   The QSSA. The rationale behind the QSSA [22] is that, after a rapid transient phase, the intermediate X reaches a quasi-stationary state in which its concentration does not change appreciably with time. Applying this assumption to Eq. (5.8) (i.e., $dx/dt = 0$) gives the concentration of the intermediate complex as

$$x_{qss} = \frac{e_0 s}{K_m + s},$$

(5.9)

where $K_m = (k_{-1} + k_2)/k_1$ is the well-known Michaelis constant. Substituting x_{qss} into the differential equation for the substrate (Eq. (5.7)) gives the rate law

$$\frac{ds}{dt} = \frac{-k_2 e_0 s}{K_m + s} = \frac{-v_m s}{K_m + s},$$

(5.10)

which is the well-known Michaelis–Menten equation, where v_m is the maximum reaction rate (or reaction velocity).

   Initially, the QSSA was justified based on physical intuition, but justification for its applicability is actually found within the theory of singular perturbations [19]. Equation (5.10) can be shown to be the first term in an asymptotic series solution derived from singular perturbation theory [41, 78]; see the review in [88].

   The QEA. Here, one assumes that the binding step quickly reaches a quasi-equilibrium state [43, 80], where

$$\frac{se}{x} = \frac{s(e_0 - x)}{x} = \frac{k_{-1}}{k_1} = K_d, \quad \text{or} \quad x_{qe} = \frac{e_0 s}{K_d + s}$$

(5.11)

holds. K_d is the disassociation equilibrium constant. Note the similarity to Eq. (5.9). Hence, one obtains the rate law

$$\frac{dp}{dt} = \frac{k_2 e_0 s}{K_d + s}$$

(5.12)

by using Eq. (5.11) in the differential equation for the product P.

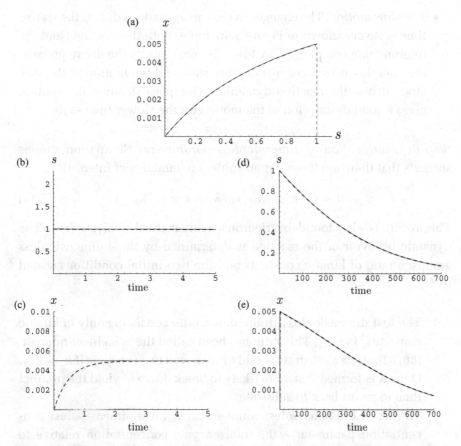

Figure 5.4 The transient response of the Michaelis–Menten reaction mechanism, for $k_2 = k_{-1}$, $100e_0 = K_m$, and $s_0 = K_m$. (a) The phase portrait. (b, c) The fast transients. (d, e) The slow transients. The solid and dashed lines represent the quasi-steady state and the full numerical solution respectively.

Step 5: Numerical solutions The full dynamic description of the kinetics of the reaction (Eqs (5.7) and (5.8)) can be obtained by direct numerical integration. The results are most conveniently shown on a phase portrait along with the transient response of the concentrations on both the fast and slow time scales; see Figure 5.4.

- *The phase portrait.* The phase portrait is shown in Figure 5.4a and it shows how the reaction rapidly approaches the quasi-steady state line and then moves along that line towards the equilibrium in the origin where the reaction has gone to completion.
- *The fast motion.* Figure 5.4b and c shows the changes in the concentrations during the faster time scale. The intermediate concentration exhibits a significant fast motion, while the substrate does not move far from its initial value.

- *The slow motion.* The changes in the concentrations during the slower time scale are shown in Figure 5.4d and e. Both the substrate and the intermediate complex decay towards zero. During the decay process, the complex is in a quasi-stationary state and the motion of the substrate drives the reaction dynamics. The quasi-steady-state solution gives a good description of the motion on the slower time scale.

Step 6: Identification of dimensionless parameters Simulation studies suggests that there are three dimensionless parameters of interest:

$$a = k_2/k_{-1}, \quad b = e_0/K_{\mathrm{m}}, \quad c = s_0/K_{\mathrm{m}}. \tag{5.13}$$

This result is also found by rigorous mathematical analysis [88]. The dynamic behavior of the reaction is determined by three dimensionless groups: a ratio of kinetic constants and the two initial conditions scaled to K_{m}.

1. The first dimensionless group, a, is a ratio consisting only of kinetic constants, k_2/k_{-1}. This ratio has been called the 'stickiness number' [88, 92], since a substrate is said to stick well to an enzyme if $k_2 > k_{-1}$. Once X is formed it is more likely to break down to yield the product than to revert back to substrate.
2. The second dimensionless number, e_0/K_{m}, is a dimensionless concentration parameter – the total enzyme concentration relative to the Michaelis constant. This quantity varies from one situation to another and takes particularly different values under *in vitro* and *in vivo* conditions. The enzyme concentrations used *in vitro* are several orders of magnitude lower than the K_{m} values [74, 115, 116]. *In vivo* enzyme concentrations can approach the same order of magnitude as K_{m}.
3. The third dimensionless ratio, s_0/K_{m}, is the initial condition for the substrate concentration. Typical values for this ratio *in vivo* are on the order of unity.

Comment on the criterion $e_0 \ll s_0$ Historically, the commonly accepted criterion for the applicability of the QSSA is that the initial concentration of the enzyme must be much smaller than that of the substrate. The actual criterion is $e_0 \ll K_{\mathrm{m}}$, or $b \ll 1$ [88]. Figure 5.5 shows the reaction dynamics for $100e_0 = K_{\mathrm{m}}$, $e_0 = s_0$, $k_2 = k_{-1}$, which is analogous to Figure 5.4, except the initial substrate concentration is now 100 times smaller than K_{m}. In other words, we have $e_0 = s_0 \ll K_{\mathrm{m}}$ and, as demonstrated in Figure 5.7, the QSSA is applicable.

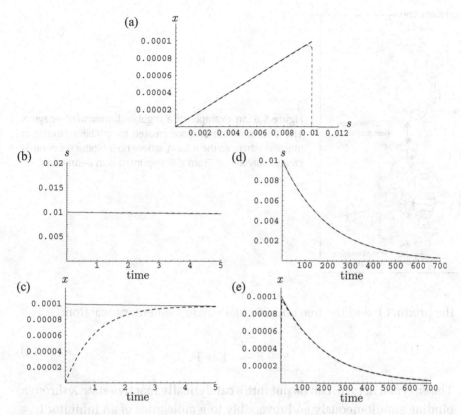

Figure 5.5 The transient response of the Michaelis–Menten reaction mechanism, for $k_2 = k_{-1}$, $100e_0 = K_m$, and $100s_0 = K_m$. (a) The phase portrait. (b, c) The fast transients. (d, e) The slow transients. The solid and dashed lines represent the quasi-steady state and the full numerical solution respectively.

5.4   Hill kinetics for enzyme regulation

Regulated enzymes   Enzyme activity is regulated by the binding of small molecules to the enzyme resulting in an altered enzymatic activity. Such binding can inhibit or activate the catalytic activities of the enzyme. The regulation of enzymes such as regulators represents a "tug of war" between the functional states of the enzyme; see Figure 5.6. A simple extension of the oldest reaction mechanisms for ligand binding to oligomeric protein, i.e., oxygen binding to hemoglobin, is commonly used to obtain simple rate laws for regulated enzymes [46].

The reaction mechanism   The Hill reaction mechanism is based on two reactions: a catalytic conversion and the sequestration of the enzyme in an inactive form. It assumes that the catalyzed reaction is an irreversible bimolecular reaction between the substrate S and the enzyme E to form

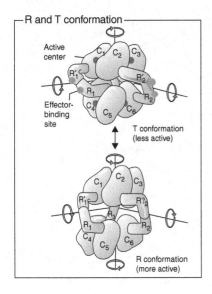

the product P and the free enzyme in a single elementary reaction:

$$S + E \xrightarrow{k} E + P.$$ (5.14)

The enzyme, in turn, can be put into a catalytically inactive state X through binding simultaneously and reversibly to ν molecules of an inhibitor I:

$$E + \nu I \underset{k_i^-}{\overset{k_i^+}{\rightleftharpoons}} X.$$ (5.15)

Numerical values for ν often exceed unity. Thus, the regulatory action of I is said to be *lumped* in the simple E to X transformation, as values of $\nu > 1$ are chemically unrealistic. Numerical values estimated from data show that the best-fit values for ν are not integers; for instance, ν is found to be around 2.3 to 2.6 for O_2 binding to hemoglobin. Section 5.5 describes more realistic reaction mechanisms of serial binding of an inhibitor to a regulated enzyme to sequester it in an inactive form.

Step 1: Dynamic mass balances   The mass action kinetic equations are

$$\frac{ds}{dt} = -v_1, \quad \frac{de}{dt} = -v_2 + v_3, \quad \frac{dp}{dt} = v_1, \quad \frac{di}{dt} = -\nu(v_2 - v_3), \quad \frac{dx}{dt} = v_2 - v_3,$$

where the reaction rates are

$$v_1 = kse, \quad v_2 = k_i^+ i^\nu e, \quad v_3 = k_i^- x.$$ (5.16)

Step 2: Finding the time invariants We define $\mathbf{x} = (s, e, p, i, x)$ and $\mathbf{v} = (ks, k_i^+ i^\nu e, k_i^- x)$. The stoichiometric matrix is then

$$
\mathbf{S} = \begin{pmatrix} -1 & 0 & 0 \\ 0 & -1 & 1 \\ 1 & 0 & 0 \\ 0 & -\nu & \nu \\ 0 & 1 & -1 \end{pmatrix} \tag{5.17}
$$

and is of rank 2. The conservation quantities are a balance on the substrate, the enzyme, and the inhibitor:

$$
s_0 = s + p, \quad e_0 = e + x, \quad i_0 = i + \nu x. \tag{5.18}
$$

Step 3: Reducing the dynamic description We need two differential equations to simulate the dynamic response and then the remaining three variables can be computed from the conservation relationships. We can choose the substrate, s, and the concentration of the enzyme, e:

$$
\frac{ds}{dt} = kse, \quad \frac{de}{dt} = -k_i^+ i^\nu e + k_i^- x; \tag{5.19}
$$

then p, x, and i are computed from Eq. (5.18).

Step 4: Applying simplifying kinetic assumptions If we assume that the binding of the inhibitor is fast, so that a quasi-equilibrium forms for the reaction of Eq. (5.15), we have

$$
v_2 = v_3; \quad \text{thus} \quad x = (k_i^+/k_i^-)i^\nu e = (i/K_i)^\nu e \quad \text{and} \quad \frac{de}{dt} = \frac{dx}{dt} = \frac{di}{dt} = 0, \tag{5.20}
$$

where K_i is a "per-site" dissociation constant for Eq. (5.15). The enzyme is in one of two states, so that we have the mass balance

$$
e_0 = e + x = [1 + (i/K_i)^\nu]e \quad \text{or} \quad e(i) = \frac{e_0}{1 + (i/K_i)^\nu}, \tag{5.21}
$$

where e_0 is the total concentration of the enzyme. Using the mass balance and the QEA gives the flux through the regulated reaction as

$$
v(i) = ke(i)s = \frac{ke_0 s}{1 + (i/K_i)^\nu} = \frac{v_m}{1 + (i/K_i)^\nu}, \tag{5.22}
$$

with $v_m = ke_0 s$. The Hill model has three parameters: (1) ν, the *degree of cooperativity*, (2) K_i, the dissociation constant for the inhibitor and, (3) v_m, the maximum reaction rate or the capacity of the enzyme.

We note that

$$
f_e = \frac{e(i)}{e_0} = \frac{1}{1 + (i/K_i)^\nu} \tag{5.23}
$$

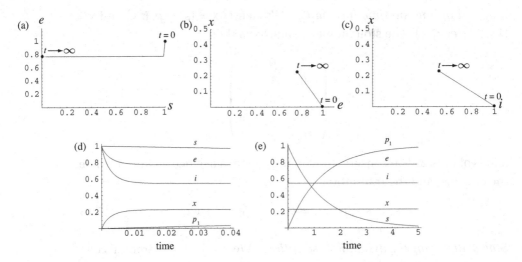

Figure 5.7 The transient response of the Hill reaction mechanism, for $k_i^+ = k_i^- = 100$, $k = 1$, $v = 2$, $x_0 = 0$, and $e_0 = s_0 = i_0 = 1$. (a) The phase portraits of s and e. (b) The phase portraits of e and x. (c) The phase portraits of i and x. (d) The fast transients. (e) The slow transients.

represents the fraction of the enzyme that is in the active state. Note that $f_e < 1$ for any finite concentration of the inhibitor.

Step 5: Numerical solutions The dynamic response of the Hill reaction mechanism is shown in Figure 5.7. The trace in the s versus e phase portrait is L-shaped, showing a rapid initial equilibration of the enzyme to the inhibitor (the vertical line), followed by the slower conversion of the product (the horizontal line). These two reactions are naturally (stoichiometrically) decoupled and separated in time for the numerical values of the kinetic constants used.

The phase portraits for e versus x and i versus x are straight lines as given by the conservation equations in (5.18); see Figure 5.7b and c. The two phase transient responses in Figure 5.7d and e show the rapid equilibration of the enzyme and the slow conversion of substrate respectively. Under these parameter conditions, the QEA should give good results.

Step 6: Estimating key parameters There are two features of the Hill rate law that are of interest:

(1) *Applicability of the QEA.* Given the fact that the two reactions have characteristic time scales, their relative magnitude is of key concern when it comes to the justification of the QEA:

$$a = \frac{\text{characteristic binding time of the inhibitor}}{\text{characteristic turnover time of the substrate}} = \frac{k}{k_i^+}. \qquad (5.24)$$

Table 5.1 The values of the function $N(v)$ and i^*/K_I at the inflection point

v	2	3	4	5
$N(v)$	0.65	0.84	1.07	1.30
i^*/K_I	0.58	0.79	0.88	0.92

If a is much smaller than unity, we would expect the QEA to be valid. In Figure 5.7, a is 0.01.

(2) *Regulatory characteristics.* The Hill rate law has a sigmoidal shape with sensitivity of the reaction rate to the end product concentration as

$$v_i = \frac{\partial v}{\partial i} = \frac{-v v_m}{i} \frac{(i/K_i)^v}{[1+(i/K_i)^v]^2}, \tag{5.25}$$

which has a maximum

$$v_i^* = -\frac{v_m}{K_i} N(v), \quad \text{where} \quad N(v) = \frac{1}{4v}(v-1)^{1-1/v}(v+1)^{1+1/v}, \tag{5.26}$$

at the inflection point

$$i^* = K_i \left(\frac{v-1}{v+1}\right)^{1/v}. \tag{5.27}$$

For plausible values of v, the function $N(v)$ is on the order of unity (Table 5.1), and hence the maximum sensitivity v_i^* is on the order of $(-v_m/K_i)$. The ratio K_i/v_m can be interpreted as a time constant characterizing the inhibition process:

$$t_i = \frac{K_i}{v_m} = \frac{\text{concentration}}{\text{concentration/time}}. \tag{5.28}$$

This estimate represents an upper bound, since the steady-state concentration of i can be different from i^*. The turnover of the substrate happens on a time scale defined by the rate constant $t_s = 1/k$. Thus, a key dimensionless property is

$$b = \frac{t_s}{t_i} = \frac{1/k}{K_i/v_m} = \frac{v_m}{kK_i} = \frac{e_t}{K_i}. \tag{5.29}$$

Therefore, the dimensionless parameter b can be interpreted as a ratio of time constants or as a ratio of concentration ranges.

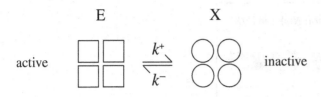

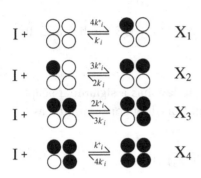

Figure 5.8 The reaction mechanisms for the symmetry model. The enzyme has four binding sites for the inhibitor.

5.5   The symmetry model

The regulatory molecules are often chemically quite different than the substrate molecule. Thus, they often have a different binding site on the protein molecule than the catalytic site. This is called an *allosteric site*. One of the earliest enzyme kinetic models that accounted for allosterism was the symmetry model [82], named after certain assumed symmetry properties of the subunits of the enzyme. It is a mechanistically realistic description of regulatory enzymes. An example of a multimeric regulatory enzyme is given in Figure 5.6.

The reaction mechanism   The main chemical conversion in the symmetry model is as before and is described by Eq. (5.14). The symmetry model postulates that the regulated enzyme lies naturally in two forms, E and X, and is converted between the two states simply as

$$E \underset{k^-}{\overset{k^+}{\rightleftharpoons}} X \tag{5.30}$$

The equilibrium constant for this reaction,

$$L = k^+/k^- = x/e, \tag{5.31}$$

has a special name, the *allosteric constant*. Then ν molecules of an inhibitor I can bind sequentially to X as

$$
\begin{aligned}
X + I &\underset{k_i^-}{\overset{\nu k_i^+}{\rightleftharpoons}} X_1 \\[2mm]
X_1 + I &\underset{2k_i^-}{\overset{(\nu-1)k_i^+}{\rightleftharpoons}} X_2 \\[2mm]
&\vdots \qquad\quad \vdots \\[2mm]
X_{\nu-1} + I &\underset{\nu k_i^-}{\overset{k_i^+}{\rightleftharpoons}} X_\nu,
\end{aligned} \tag{5.32}
$$

where the binding steps have the same dissociation constant, $K_i = k_i^-/k_i^+$. We will discuss the most common case of a tetramer here, i.e., $\nu = 4$; see Figure 5.8.

Step 1: Dynamic mass balances The conversion rate of the substrate is

$$
v = kse, \tag{5.33}
$$

whereas the enzyme sequestration is characterized by the reaction rates

$$
v_1 = k^+ e \qquad v_2 = k^- x \qquad v_3 = 4k_i^+ x i, \tag{5.34}
$$

$$
v_4 = k_i^- x_1 \qquad v_5 = 3k_i^+ x_1 i \qquad v_6 = 2k_i^- x_2, \tag{5.35}
$$

$$
v_7 = 2k_i^+ i x_2 \qquad v_8 = 3k_i^- x_3 \qquad v_9 = k_i^+ i x_3, \tag{5.36}
$$

$$
v_{10} = 4k_i^- x_4. \tag{5.37}
$$

The dynamic mass balances on the various states of the enzyme are:

$$
\frac{de}{dt} = -v_1 + v_2, \tag{5.38}
$$

$$
\frac{dx}{dt} = v_1 - v_2 - v_3 + v_4, \tag{5.39}
$$

$$
\frac{di}{dt} = -v_3 + v_4 - v_5 + v_6 - v_7 + v_8 - v_9 + v_{10}, \tag{5.40}
$$

$$
\frac{dx_1}{dt} = v_3 - v_4 - v_5 + v_6, \tag{5.41}
$$

$$
\frac{dx_2}{dt} = v_5 - v_6 - v_7 + v_8, \tag{5.42}
$$

$$
\frac{dx_3}{dt} = v_7 - v_8 - v_9 + v_{10}, \tag{5.43}
$$

$$
\frac{dx_4}{dt} = v_9 - v_{10}. \tag{5.44}
$$

Step 2: Finding the time invariants   The stoichiometric matrix for $\mathbf{x} =$ $(e, x, i, x_1, x_2, x_3, x_4)$ is the 7×10 matrix

$$S = \begin{pmatrix} -1 & 1 & 0 & 0 & 0 & 0 & 0 & 0 & 0 & 0 \\ 1 & -1 & -1 & 1 & 0 & 0 & 0 & 0 & 0 & 0 \\ 0 & 0 & -1 & 1 & -1 & 1 & -1 & 1 & -1 & 1 \\ 0 & 0 & 1 & -1 & -1 & 1 & 0 & 0 & 0 & 0 \\ 0 & 0 & 0 & 0 & 1 & -1 & -1 & 1 & 0 & 0 \\ 0 & 0 & 0 & 0 & 0 & 0 & 1 & -1 & -1 & 1 \\ 0 & 0 & 0 & 0 & 0 & 0 & 0 & 0 & 1 & -1 \end{pmatrix} \qquad (5.45)$$

that has a rank of 5. Thus, there are two conservation relationships: for the enzyme, $e_0 = e + x + x_1 + x_2 + x_3 + x_4$, and for the inhibitor, $i_0 = i + x_1 + 2x_2 + 3x_3 + 4x_4$. If the dynamic mass balances on the substrate and product are taken into account, then a third conservation $s_0 = s + p$ appears.

Step 3: Reducing the dynamic description   We leave it to the reader to pick two dynamic variables from the full kinetic model as the dependent variables and then eliminate them from the dynamic description using the conservation relationships. The impetus for doing so algebraically becomes smaller as the number of differential equations grows. Most standard software packages will integrate a dynamically redundant set of differential equations, and such substitution is not necessary to obtain the numerical solutions.

Step 4: Using simplifying kinetic assumptions to derive a rate law   The serial binding of an inhibitor to X that has four binding sites is shown in Figure 5.8. The derivation of the rate law is comprised of four basic steps.

1. Mass balance on enzyme:

$$e_0 = e + x + x_1 + x_2 + x_3 + x_4. \qquad (5.46)$$

2. QEA for binding steps:

$$\begin{aligned} 4k_i^+ ix &= k_i^- x_1 &\Rightarrow x_1 &= 4x(i/K_i) = 4x\,(i/K_i), \\ 3k_i^+ ix_1 &= 2k_i^- x_2 \Rightarrow x_2 &= \tfrac{3}{2}x_1(i/K_i) = 6x(i/K_i)^2, \\ 2k_i^+ ix_2 &= 3k_i^- x_3 \Rightarrow x_3 &= \tfrac{2}{3}x_2(i/K_i) = 4x(i/K_i)^3, \\ k_i^+ ix_3 &= 4k_i^- x_4 &\Rightarrow x_4 &= \tfrac{1}{4}x_3(i/K_i) = x(i/K_i)^4. \end{aligned} \qquad (5.47)$$

3. Combine 1 and 2:

$$\begin{aligned} e_0 &= e + x + 4x(i/K_i) + 6x(i/K_i)^2 + 4x(i/K_i)^3 + x(i/K_i)^4 \\ &= e + x[1 + (i/K_i)]^4 \qquad \text{where } \ x = Le \qquad (5.48) \\ &= e\{1 + L[1 + (i/K_i)]\}^4 \end{aligned}$$

4. Form the rate law: the reaction rate is given by $v = kse$. We can rewrite the last part of Eq. (5.48) as

$$e = \frac{e_0}{1 + L(1 + i/K_i)^4},$$ (5.49)

leading to the rate law

$$v(s, i) = \frac{ke_0 s}{1 + L(1 + i/K_i)^4}.$$ (5.50)

This rate law generalizes to

$$v(s, i) = \frac{ke_0 s}{1 + L(1 + i/K_i)^\nu} = \frac{v_m}{1 + L(1 + i/K_i)^\nu}$$ (5.51)

for any ν. The reader can find the same key dimensionless groups as for the Hill rate law. Note again the fraction

$$f_e = \frac{e}{e_0} = \frac{1}{1 + L(1 + i/K_i)^\nu},$$ (5.52)

which describes what fraction of the enzyme is in the catalytically active state.

Step 5: Numerical solutions These equations can be simulated. Typically, the conformational changes between E and X are fast, as are the inhibitor binding steps relative to the catalysis rate. Numerical simulations were carried out for this situation and the results are plotted in Figure 5.9.

- Figure 5.9a shows how the substrate–enzyme phase portrait is L-shaped, showing that the sequestration of the enzyme in the inhibited form (the vertical line) is faster than the conversion of the substrate (the horizontal line).
- Figure 5.9b shows the redistribution of the total enzyme among the active and inactive forms; that is, e versus $x + x_1 + x_2 + x_3 + x_4$. The fraction of the enzyme in the inactive form is about 0.29.
- Figure 5.9c shows the redistribution of the inhibitor between the free and bound form; i versus $x_1 + 2x_2 + 3x_3 + 4x_4$. This panel shows that the fraction of inhibitor that is bound is high: 0.70.
- Finally, Figure 5.9d and e show the transient changes in the concentrations and pools on the fast and slow time scales respectively. Note that two natural aggregation variables appear: the total enzyme in the inactive form and the total number of inhibitor molecules bound to the enzyme.

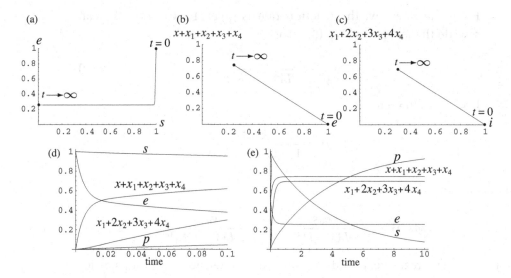

Figure 5.9 The transient response of the symmetry model, for $k^+ = k^- = 100$, $k_i^+ = k_i^- = 100$, $k = 1$, $v = 4$, $x_0 = x_{1,0} = x_{2,0} = x_{3,0} = x_{4,0} = 0$, and $e_0 = s_0 = i_0 = 1$. (a) The phase portraits of s and e. (b) The phase portraits of e and $x + x_1 + x_2 + x_3 + x_4$. (c) The phase portraits of i and $x_1 + 2x_2 + 3x_3 + 4x_4$. (d) The fast transients. (e) The slow transients.

5.6   Scaling dynamic descriptions

The analysis of simple equations requires the "proper frame of mind." In step 6 of the process of formulating rate laws, this notion is translated into quantitative measures. We need to scale the variables with respect to intrinsic reference scales and thereby cast our mathematical descriptions into appropriate coordinate systems. All parameters then aggregate into dimensionless property ratios that, if properly interpreted, have a clear physical significance.

The scaling process   The examples above illustrate the decisive role of time constants and their use to analyze simple situations and to elucidate intrinsic reference scales. Identification of unimportant terms is sometimes more difficult, and familiarity with a formal scaling procedure is useful. This procedure basically consists of four steps:

1. Identify logical reference scales. This step is perhaps the most difficult. It relies partly on physical intuition, and the use of time constants is surprisingly powerful even when analyzing steady situations.
2. Introduce reference scales into the equations and make the variables dimensionless.

3. Collect the parameters into dimensionless property ratios. The number of dimensionless parameters is always the same and it is given by the well-known Buckingham Pi theorem.[1]

4. Interpret the results. The dimensionless groups that appear can normally be interpreted as ratios of the time constants, such as those discussed above.

Scaling of equations is typically only practiced for small models and for analysis purposes only. Numerical simulations of complex models are essentially always performed with absolute values of the variables.

The importance of intrinsic reference scales The process by which the equations are made dimensionless is not unique. The "correct" way of putting the equations into a dimensionless form, where judgments of relative orders of magnitude can be made, is called *scaling*. The scaling process is defined by Lin and Segel [72] as

> . . . select intrinsic reference quantities so that each term in the dimensional equations transforms into a product of a constant dimensional factor which closely estimates the term's order of magnitude and a dimensionless factor of unit order of magnitude.

In other words, if one has an equation which is a sum of terms T_i as

$$T_1 + T_2 + \cdots = 0, \tag{5.53}$$

one tries to scale the *variables* involved so that they are of unit order of magnitude or

$$t_i = \frac{\text{variable}_i}{(\text{intrinsic reference scale})_i} \approx \text{unit order of magnitude.} \tag{5.54}$$

Introducing these dimensionless variables into Eq. (5.53) results in the dimensionless form

$$\pi_1 t_1 + \pi_2 t_2 + \cdots = 0, \tag{5.55}$$

where the dimensionless multipliers π_i are the dimensionless groups and they will indicate the order of magnitude of the product $\pi_i t_i$. Once the equations are in this form, order-of-magnitude judgements can be made based on the dimensionless groups.

---

[1] This theorem states that the number of dimensionless groups in a given problem is equal to the number of dimensioned parameters that appear in the original set of equations, minus the number of fundamental dimensions involved.

5.7   Summary

➤ Enzymes are highly specialized catalysts that can dramatically accelerate the rates of biochemical reactions.

➤ Reaction mechanisms are formulated for the chemical conversions carried out by enzymes in terms of elementary reactions.

➤ Rate laws for enzyme reaction mechanisms are derived based on simplifying assumptions.

➤ Two simplifying assumptions are commonly used: the QSSA and the QEA.

➤ The validity of the simplifying assumptions can be determined using scaling of the equations followed by mathematical and numerical analysis.

➤ A number of rate laws have been developed for enzyme catalysis and for the regulation of enzymes. Only three reaction mechanisms were described in this chapter.

Open systems

All of the examples in the previous two chapters were closed systems. The ultimate state of closed, chemically reacting systems is chemical equilibrium. Living systems are characterized by mass and energy flows across their boundaries that keep them away from equilibrium. We now discuss *open systems*, which allow molecules to enter and leave. Open systems ultimately reach a steady state that is often close to a *homeostatic* state of interest.

6.1   Basic concepts

There are several fundamental concepts that need to be understood when one considers open systems. We discuss the more significant ones in this section.

The system boundary   Implicit in the term "open system" is the notion of an inside and an outside, the division of the world into two domains. The separation between the two is the *system boundary*, which thus defines what is inside a system, and belongs to it, and what is outside.

The definition of a boundary can be physical. An example of a physical boundary may be the cell wall, which clearly defines what is inside and what is outside. Similarly, the outer membrane of the mitochondria can serve as a clearly defined physical system boundary. Thus, systems can have hard, immovable boundaries or soft, flexible ones. In the latter case, the volume of the system may be changing.

The definition of a boundary can also be virtual. For instance, we can define a pathway, such as the tricarboxylic acid cycle, as a system, or the amino acid biosynthetic pathways in a cell as a system. In both cases, we

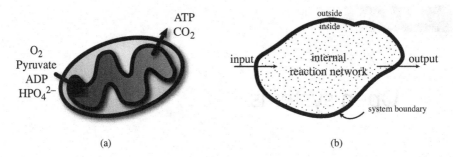

Figure 6.1 Open systems. (a) Example production of ATP from pyruvate by mitochondria (prepared by Nathan Lewis). (b) Basic definitions associated with an open system.

might draw dashed lines around it on the metabolic map to indicate the system boundary.

Crossing the boundary: inputs and outputs   Once a system boundary has been established, we can identify and define interactions across it. They are the *inputs* and *outputs* of the system. In most cases, we will be considering mass flows in and out of a system; molecules coming and going. Such flows are normally amenable to experimental determination.

There may be other quantities crossing the system boundary. For instance, if we are considering photosynthesis, photons are crossing the system boundary, providing a net influx of energy. Forces can also act on the system boundary. For instance, if the number of molecules coming into and leaving a system through a coupled transport mechanism is not balanced, then osmotic pressure may be generated across the system boundary. Cells have sodium–potassium pumps that displace an uneven number of two types of cation to deal with osmotic imbalances.

Perturbations and boundaries   In most analyses of living systems, we are concerned about changes in the environment. The temperature may change, substrate availability may change, and so on. Such changes in the environment can be thought of as forcing functions to which the living system responds. These are one-way interactions, from the environment to the system. In most cases, we consider the system to be small (relative to the environment) and that the environment buffers the activities of the system. For instance, the environment may provide an infinite sink for cellular waste products.

In other cases, the activities of the system influence the state of the environment. In batch fermentation, for example, the metabolic activities will substantially change the chemical composition of the medium during the course of the fermentation. On a global scale, we are now becoming more concerned about the impact that human activities have on the climate and the larger environment of human socio-economic activities. In such cases, there are two-way interactions between the system and the environment that need to be described.

Inside the boundary: the internal network The definition of a system boundary determines what is inside. Once we know what is inside the system, we can determine the network of chemical transformations that takes place. This network has topology and its links have kinetic properties.

In most cases, it is hard to measure the internal state of a system. There typically are only a few noninvasive measurements available. Occasionally, there are probes that we can use to observe the internal activities. Tracers, such as ^{13}C atoms strategically placed in substrates, can be used to trace input to output and allow us to determine pieces of the internal state of a system. Tearing a system apart to enumerate its components is, of course, possible, but the system is destroyed in the process. Thus, we are most often in the situation where we cannot fully experimentally determine the internal state of a system, and may have to be satisfied with only partial knowledge.

From networks to system models Mathematical models may help us to simulate or estimate the internal state of the system. Sometimes we are able to bracket its state based on the inputs and outputs and our knowledge of the internal network.

Full dynamic simulation requires extensive knowledge about the internal properties of a system. Detailed models that describe the dynamic state of a system require the definition of the system boundary, the inputs and outputs, the structure of the internal network, and the kinetic properties of the links in the network. We can then simulate the response of the system to various perturbations, such as changes in the environmental conditions.

The functional state Once a system model has been formulated, it can be used to compute the functional state of the system for a given set of conditions. Closed systems will eventually go to chemical equilibrium. However, open systems are fundamentally different. Owing to continuous interactions with the environment, they have internal states that are either dynamic or steady. A system that has fast internal time constants relative to changes in the environment will reach a steady state, or a quasi-steady state. For a biological system, such functional states are called *homeostatic states*. Such states are maintained through energy dissipation. The flow of energy in has to exceed the energy leaving the system. This difference allows living systems to reach a functional homeostatic state.

6.2 Reversible reaction in an open environment

In an open environment, the simple reaction of Eq. (4.1) has an inflow (b_1) of x_1 and an outflow (b_2) of x_2:

$$\overset{b_1}{\rightarrow} x_1 \underset{v_{-1}}{\overset{v_1}{\rightleftharpoons}} x_2 \overset{b_2}{\rightarrow} . \tag{6.1}$$

The input is fixed by the environment and there is a first-order rate for the product to leave the system, and thus we have

$$b_1 = \text{constant} \quad \text{and} \quad b_2 = k_2 x_2.$$

This defines the system boundary, and the inputs and outputs.

The stoichiometric matrix is

$$\mathbf{S} = \begin{pmatrix} 1 & -1 & 1 & 0 \\ 0 & 1 & -1 & -1 \end{pmatrix}, \tag{6.2}$$

where $\mathbf{x} = (x_1, x_2)$ and $\mathbf{v} = (b_1, k_1 x_1, k_{-1} x_2, k_2 x_2)$. The stoichiometric matrix is of rank 2 and is thus a two-dimensional dynamic system. The differential equations that will need to be solved are

$$\frac{dx_1}{dt} = b_1 - k_1 x_1 + k_{-1} x_2 \quad \text{and} \quad \frac{dx_2}{dt} = k_1 x_1 - k_{-1} x_2 - k_2 x_2. \tag{6.3}$$

There are no conservation quantities.

The steady states   There are three properties of the steady state of interest. We can find first the steady-state fluxes and second the steady-state concentrations. Third, we can determine the difference between the steady state and the equilibrium state of this open system.

(1) *Steady-state fluxes.* The steady state of the fluxes is given by

$$\mathbf{S}\mathbf{v}_{ss} = \mathbf{0}, \tag{6.4}$$

and thus $\mathbf{v}_{ss}$ resides in the null space of $\mathbf{S}$. For this matrix, the null space is two-dimensional: $\text{Dim}(\text{Null}(\mathbf{S})) = n - r = 4 - 2 = 2$, where $n = 4$ is the number of fluxes and the rank is $r = 2$. The null space is spanned by two pathway vectors: $(1,1,0,1)$ and $(0,1,1,0)$. The former is the path through the system and the latter corresponds to the reversible reaction. These are known as type 1 and type 3 extreme pathways respectively [89]. All steady-state flux states of the system are a non-negative combination of these two vectors:

$$\mathbf{v}_{ss} = (b_1, k_1 x_{1,ss}, k_{-1} x_{2,ss}, k_2 x_{2,ss}) \tag{6.5}$$

$$= a(1, 1, 0, 1) + b(0, 1, 1, 0), \quad a \geq 0, \ b \geq 0. \tag{6.6}$$

(2) *Steady-state concentrations.* The concentrations in the steady state can be evaluated from

$$0 = b_1 - k_1 x_1 + k_{-1} x_2, \quad 0 = k_1 x_1 - k_{-1} x_2 - k_2 x_2. \tag{6.7}$$

If we add these equations we find that $x_{2,ss} = b_1/k_2$. This concentration can then be substituted into either of the two equations to show that $x_{1,ss} = (1 + k_{-1}/k_2)(b_1/k_1)$. Thus, the steady-state concentration vector is

$$\mathbf{x}_{ss} = \begin{pmatrix} x_{1,ss} \\ x_{2,ss} \end{pmatrix} = \begin{pmatrix} b_1 \\ k_2 \end{pmatrix} \begin{pmatrix} \frac{k_2 + k_{-1}}{k_1} \\ 1 \end{pmatrix}. \tag{6.8}$$

These steady-state concentrations can be substituted into the steady-state flux vector to get

$$\mathbf{v}_{ss} = b_1 (1, 1 + (k_{-1}/k_2), (k_{-1}/k_2), 1) \tag{6.9}$$

$$= a(1, 1, 0, 1) + b(0, 1, 1, 0), \quad a = b_1, \ b = \begin{pmatrix} b_1 k_{-1} \\ k_2 \end{pmatrix}. \tag{6.10}$$

Therefore, the steady-state flux distribution is a summation of the straight-through pathway and the discounted flux through the reversible reaction. The key quantity is k_{-1}/k_2, which measures the relative rate of x_2 reacting back to form x_1 versus the rate at which it leaves the system.

(3) *The distance from the equilibrium state.* The difference between the steady state and the equilibrium state can be measured by

$$\frac{x_{2,ss}/x_{1,ss}}{x_{2,eq}/x_{1,eq}} = \frac{1}{1 + k_2/k_{-1}} \xrightarrow{k_2 \ll k_{-1}} 1. \tag{6.11}$$

Thus, when $k_2 \ll k_{-1}$, the steady state approaches the equilibrium state (recall that $(x_{2,ss}/x_{1,ss})/(x_{2,eq}/x_{1,eq}) = \Gamma/K_{eq}$, Section 2.2). If the exchange with the environment is slow relative to the internal reaction rates, then the internal system approaches that of an equilibrium state.

Dynamic states The dynamic states are computed from the dynamic mass balances for a given condition. We are interested in two dynamic states: the approach to the steady state and the response to a change in the input flux b_1.

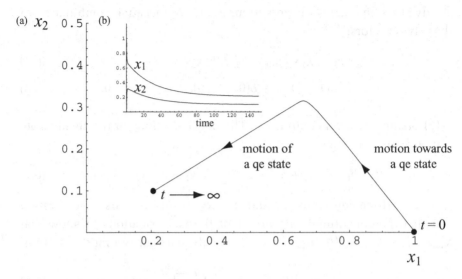

Figure 6.2 The concentration–time profiles for the reaction system $\rightarrow x_1 \rightleftharpoons x_2 \rightarrow$ for $k_1 = 1$, $k_{-1} = 2$, $k_2 = 0.1$, $x_1(0) = 1$, $x_2(0) = 0$, and $b_1 = 0.01$. (a) The phase portrait of x_1 and x_2. (b) The concentrations as a function of time.

(1) *A two-phase transient response.* Simulation of this system with an input rate of $b_1 = 0.01$ and a slow removal rate, $k_2 = 0.1$, from an initial state of $x_1(0) = 1.0$ and $x_2(0) = 0.0$ is shown in Figure 6.2. There are two discernible time scales: a rapid motion of the equilibrating reaction and a slow removal of x_2 from the system; Figure 6.2b. A phase portrait shows a rapid movement along a line with a negative slope of 1, showing the existence of a conservation quantity $(x_1 + x_2)$ on a fast time scale, followed by a slow motion down a quasi-equilibrium line with a slope of 1/2 to a steady-state point. Note the difference from Figure 4.1c.

The same pools can be formed as defined in Eq. (4.9):

$$\begin{pmatrix} p_1 \\ p_2 \end{pmatrix} = \begin{pmatrix} 1 & -1/K_1 \\ 1 & 1 \end{pmatrix} \begin{pmatrix} x_1 \\ x_2 \end{pmatrix}. \tag{6.12}$$

This matrix can be used to post-process the concentrations and the results can be graphed; see Figure 6.3. The pool transformation leads to dynamic decoupling and is clearly illustrated in this figure. The disequilibrium pool relaxes very quickly, while the conservation pool moves slowly; see Figure 6.3b. The phase portrait formed by the pools thus has an L shape (Figure 6.3a), illustrating the dynamic decoupling. Note that:

- there is first a vertical motion where $x_1 + x_2$ is essentially a constant, followed by a slow horizontal motion where the disequilibrium variable $x_1 - x_2/K_1$ is a constant;

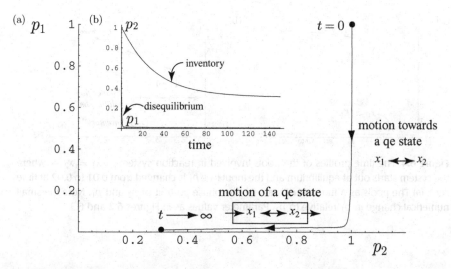

Figure 6.3 The time profiles of the pools involved in reaction system $\rightarrow x_1 \rightleftharpoons x_2 \rightarrow$ for the same conditions as in Figure 6.2. (a) The phase portrait of p_2 and p_1. (b) The pools as a function of time.

- these two separate motions correspond to forming the pathways found in the steady state; and
- the disequilibrium aggregate is not zero, as it is forced away from equilibrium by the input and settling in a steady state.

(2) *External disturbance.* The previous simulation represents a biologically unrealistic situation. An internal concentration cannot suddenly deviate from its value independent of what else happens in the system. A much more realistic situation is one where we start out at a steady state and an environmental change is observed. In our case, the only environmental parameter is b_1.

In Figure 6.4 we change the input flux b_1 from 0.01 to 0.02 at time zero, when the system is initially in a steady state (the endpoint in Figure 6.3). We make three observations:

- We see that the fast motion is not activated.
- The "inventory" or the pool of $x_1 + x_2$ moves from one steady state to another. It increases, as the forcing function was stepped up.
- The "distance from equilibrium"

$$x_{1,\text{ss}} - \frac{x_{2,\text{ss}}}{K_1} = x_{1,\text{ss}} - \frac{x_{2,\text{ss}}}{x_{2,\text{eq}}/x_{1,\text{eq}}} = x_{1,\text{ss}}\left(1 - \frac{\Gamma}{K_1}\right) \qquad (6.13)$$

is close to zero in both steady states. The higher throughput, however, does push the system farther from equilibrium.

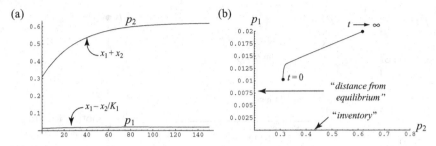

Figure 6.4 The time profiles of the pools involved in reaction system $\rightarrow x_1 \rightleftharpoons x_2 \rightarrow$ where the system starts out at equilibrium and the input rate b_1 is changed from 0.01 to 0.02 at time zero. (a) The pools as a function of time. (b) The phase portrait of p_2 and p_1. Note the small numerical change in p_1 relative to p_2. Parameter values as in Figures 6.2 and 6.3.

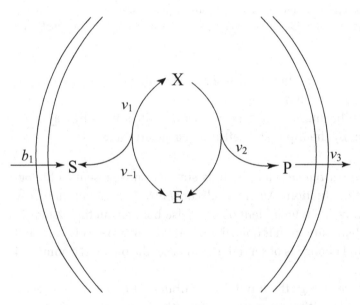

Figure 6.5 The Michaelis–Menten reaction mechanisms in an open setting. The substrate and the product enter and leave the system, while the enzyme stays inside.

6.3   Michaelis–Menten kinetics in an open environment

We now consider the case when the Michaelis–Menten reaction mechanism operates in an open environment (Figure 6.5). The substrate enters the system and the product leaves. The enzyme stays internal to the system.

Dynamic description The mass action kinetic model is

$$
\begin{aligned}
ds/dt &= b_1 - k_1 es + k_{-1}x & s(t=0) &= s_0, \\
dx/dt &= k_1 es - (k_{-1} + k_2)x & x(t=0) &= x_0, \\
de/dt &= -k_1 es + (k_{-1} + k_2)x & e(t=0) &= e_0, \\
dp/dt &= k_2 x - k_3 p & p(t=0) &= p_0.
\end{aligned}
\tag{6.14}
$$

The initial conditions would normally be the steady-state conditions. The stoichiometric matrix is

$$
\mathbf{S} = \begin{pmatrix}
1 & -1 & 1 & 0 & 0 \\
0 & -1 & 1 & 1 & 0 \\
0 & 1 & -1 & -1 & 0 \\
0 & 0 & 0 & 1 & -1
\end{pmatrix},
\tag{6.15}
$$

where $\mathbf{x} = (s, e, x, p)$ and $\mathbf{v} = (b_1, k_1 es, k_{-1}x, k_2 x, k_3 p)$.

The steady state As in the previous section, we can compute the steady-state fluxes and concentrations. We can also compute how the parameters determine the distance from equilibrium, and here we also run into an additional issue: capacity constraints that result from a conservation quantity.

(1) *The steady-state fluxes.* The rank of $\mathbf{S}$ is 3; thus, the dimension of the null space is $5 - 3 = 2$. The null space of $\mathbf{S}$ is spanned by two vectors, $(1,1,0,1,1)$ and $(0,1,1,0,0)$, that correspond to a pathway through the system and an internal reversible reaction. The steady-state fluxes are given by

$$
\mathbf{v}_{ss} = (b_1, k_1 e_{ss} s_{ss}, k_{-1}x_{ss}, k_2 x_{ss}, k_3 p_{ss})
\tag{6.16}
$$

$$
= a(1, 1, 0, 1, 1) + b(0, 1, 1, 0, 0), \quad a \geq 0, \; b \geq 0.
\tag{6.17}
$$

(2) *The steady-state concentrations.* The dimension of the left null space is $4 - 3 = 1$. The left null space has one conservation quantity $(e + x)$, which can readily be seen from the fact that the second and third rows of $\mathbf{S}$ add up to zero. The steady-state flux balances for this system are

$$
b_1 = v_1 - v_{-1} = v_2 = v_3.
\tag{6.18}
$$

Thus, the incoming flux and the kinetic parameters immediately set the concentrations for X, P, and E as

$$
x_{ss} = b_1/k_2, \quad p_{ss} = b_1/k_3, \quad \text{and} \quad e_{ss} = e_t - x_{ss} = e_t - b_1/k_2 \tag{6.19}
$$

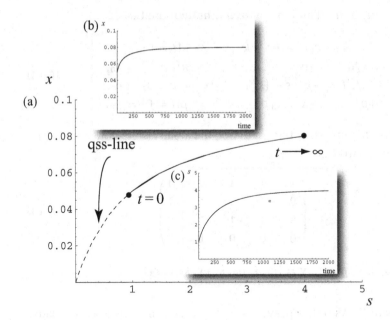

Figure 6.6 The time profiles for the transient response of the Michaelis–Menten mechanisms for $b_1 = 0.025$ changed to 0.04 at time zero. The kinetic parameters are: $e_t = 0.1$, $k_1 = 1$, $k_{-1} = 0.5$, $k_2 = 0.5$, $k_3 = 0.1$. (a) The phase portrait of s versus x. (b) The time response of x. (c) The time response of s.

and the steady-state substrate concentration can be determined. The steady-state concentration of the substrate is given by $k_1 s_{ss} e_{ss} = b_1 + k_{-1} x_{ss}$, which can be solved to give

$$s_{ss} = \left(\frac{k_2}{k_1}\right) \left(\frac{k_{-1}/k_2 + 1}{e_t k_2/b_1 - 1}\right). \tag{6.20}$$

The steady-state flux vector can now be computed.

$$\mathbf{v}_{ss} = (b_1, b_1(1 + k_{-1}/k_2), b_1 k_{-1}/k_2, b_1, b_1) \tag{6.21}$$

$$= a(1, 1, 0, 1, 1) + b(0, 1, 1, 0, 0), \quad a = b_1, \ b = b_1 k_{-1}/k_2. \tag{6.22}$$

The distance from equilibrium can now be computed as in the previous section.

(3) *Internal capacity constraints.* Since $e_{ss} \geq 0$, the maximum input is

$$b_1 \leq b_{1,\max} = k_2 e_t = v_m, \tag{6.23}$$

which is the maximum reaction rate for the Michaelis–Menten mechanism. The total amount of the enzyme and the turnover rate set this flux constraint.

Dynamic states   The response of this system to a change in the input rate starting from a steady state is of greatest interest; see Figure 6.6. For the kinetic parameters given in the figure, the steady-state concentrations of s and x are 1.0 and 0.5 respectively. The input rate is changed from 0.025 to 0.04 at time zero and the concentrations of s and x go to 4.0 and 0.8 respectively as time goes to infinity. The maximum flux rate is 0.05.

The change in the input rate triggers an internal motion that basically follows the quasi-steady state line (Figure 6.6a). Figure 6.6b and c shows that since the internal steps are rapid relative to the exchange rates, there are no rapid transients produced by a perturbation in the input rate.

If the input is increased towards v_m, the substrate concentration builds up to a very high value and most of the enzyme is found in the intermediate state. If the input rate exceeds v_m there will be no steady state, as the enzyme cannot convert the substrate to the product at the same rate as it is entering the system.

The dynamic properties of this open system can be analyzed using pool formation.

6.4   Summary

➤ Open systems eventually reach a steady state, which is different from the equilibrium state of a closed system. Such steady states can be thought of as homeostatic, living states.

➤ Living cells are open systems that continually exchange mass and energy with their environment. The continual net throughput of mass and concomitant energy dissipation is what allows steady states to form and differentiates them from equilibrium states.

➤ The relative rates of the internal network to those of the exchanges across the system boundary are important. Time-scale separation between internal and exchange processes can form. Rapid internal transients lead to pool formation.

➤ Open systems are most naturally in a steady state and respond to external stimuli. It is normally not possible to suddenly change the internal state of the system, since it is in balance with the environment.

➤ If the internal dynamics are fast, they do not excite when external stimuli are experienced, and thus accurate information about the fast kinetics may not be needed. It may be enough to know that they are "fast."

➤ Nonexchanged moieties form dynamic invariants. They can set internal capacity constraints.

## PART II

Biological characteristics

In Part I we learned how to set up simulation models and apply them to simple situations. We now turn our attention to the development of models of prototypic biological processes. To do so, we must specify numerical values for kinetic and other parameters. We thus begin by estimating reasonable numerical values of key quantities through order-of-magnitude procedures. We then examine the dynamic consequences of the stoichiometric structure of a network. Finally, regulatory mechanisms are examined to see how they change the natural mass action kinetic trends found in a network.

## CHAPTER 7

Orders of magnitude

The simulation examples in the previous chapters are conceptual. As we begin to build simulation models of realistic biological processes, we need to obtain information such as the numerical values of the parameters that appear in the dynamic mass balances. We thus go through a process of estimating the approximate numerical values of various quantities and parameters. Size, mass, chemical composition, metabolic complexity, and genetic makeup represent characteristics for which we now have extensive data available. Based on typical values for these quantities, we show how one can make useful estimates of concentrations and the dynamic features of the intracellular environment.

7.1 Cellular composition and ultra-structure

It is often stated that all biologists have two favorite organisms, *Escherichia coli* and another one. Fortunately, much data exists for *E. coli*, and we can go through parameter and variable estimation procedures using it as an example. These estimation procedures can be performed for other target organisms, cell types, and cellular processes in an analogous manner if the appropriate data are available. We organize the discussion around key questions.

The interior of a cell The typical bacterial cell, like *E. coli*, is on the order of micrometers in size (Figure 7.1a). The *E. coli* cell is a short cylinder, about 2–4 μm in length with a 0.5 to 1.5 μm diameter; Figure 7.1b. The size of the *E. coli* cell is growth-rate dependent; the faster the cell grows, the larger it is.

It has a complex intracellular environment. One can isolate and crystallize macromolecules and obtain their individual structure. However,

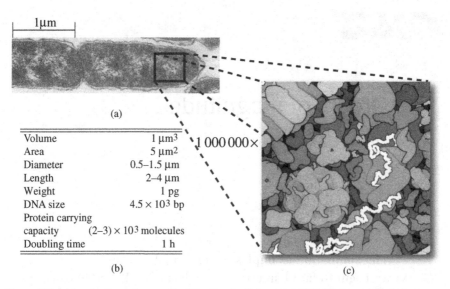

Figure 7.1 (a) An electron micrograph of the *E. coli* cell, from Ref. [50]. (b) Characteristic features of an *E. coli* cell. (c) The interior of an *E. coli* cell ©David S. Goodsell 1999.

this approach gives limited information about the configuration and location of a protein in a living functional cell. The intracellular milieu can be reconstructed from available data to yield an indirect picture of the interior of an cell. Such a reconstruction has been carried out [38]. Based on well-known chemical composition data, this image provides us with about a million-fold magnification of the interior of an *E. coli* cell; see Figure 7.1c. Examination of this picture of the interior of the *E. coli* cell is instructive:

- The intracellular environment is very crowded and represents a dense solution. Protein density in some subcellular structures can approach that found in protein crystals, and the intracellular environment is sometimes referred to as "soft glass," suggesting that it is close to a crystalline state.
- The chemical composition of this dense mixture is very complex. The majority of cellular mass is macromolecules with metabolites, the small molecular weight molecules interspersed among the macromolecules.
- In this crowded solution, the motion of the macromolecules is estimated to be 100- to even 1000-fold slower than in a dilute solution. The time it takes a 160 kDa protein to move 10 nm – a distance that corresponds approximately to the size of the protein molecule – is estimated to be 0.2 to 2 ms. Moving one cellular diameter of approximately 0.64 μm, or 640 nm, would then require 1–10 min. The motion

Table 7.1 Approximate composition of cells. The numbers given are weight percent

Component	Bacterial cell	Mammalian cell
Water	70	70
Small molecules		
Inorganic ions	1	1
Small metabolites	3	3
Macromolecules		
Protein	15	18
RNA	6	1.1
DNA	1	0.25
Phospholipids	2	3
Other lipids	0	2
Polysaccharides	2	
Total volume (μm^3)	2	4000
Relative size	1	2000

From Ref. [7].

of metabolites is expected to be significantly faster owing to their smaller size.

The overall chemical composition of a cell With these initial observations, let us take a closer look at the chemical composition of the cell. Most cells are about 70% water. It is likely that cellular functions and evolutionary design are constrained by the solvent capacity of water, and the fact that most cells are approximately 70% water suggests that all cells are close to these constraints.

The "biomass" is about 30% of cell weight. It is sometimes referred to as the dry weight of a cell, and is denoted by gDW. The 30% of the weight that is biomass is comprised of 26% macromolecules, 1% inorganic ions, and 3% low molecular weight metabolites. The basic chemical makeup of prokaryotic cells is shown in Table 7.1, and contrasted to that of a typical animal cell. The gross chemical composition is similar, except that animal cells, and eukaryotic cells in general, have a higher lipid content because of the membrane requirement for cellular compartmentalization. Approximate cellular composition is available or relatively easy to get for other cell types.

The detailed composition of E. coli The total weight of a bacterial cell is about 1 pg. The density of cells barely exceeds that of water, and cellular density is typically around 1.04 to 1.08 g/cm^3. Since the density of cells is close to unity, a cellular concentration of about 10^{12} cells per milliliter represents the packing density of *E. coli* cells. Detailed and recent data are found later in Figure 7.3.

This information provides the basis for estimating the numerical values for a number of important quantities that relate to dynamic network modeling. Having such order-of-magnitude information provides a frame of reference, allows one to develop a conceptual model of cells, evaluate the numerical outputs from models, and perform any approximation or simplification that is useful and justified based on the numbers. "Numbers count," even in biology.

Order-of-magnitude estimates   It is relatively easy to estimate the approximate order of magnitude of the numerical values of key quantities. Enrico Fermi, the famous physicist, was well known for his skills with such calculations, and they are thus known as Fermi problems. We give a couple of examples to illustrate the order-of-magnitude estimation process.

(1) *How many piano tuners are in Chicago?* This question represents a classical Fermi problem. First, we state assumptions or key numbers:

1. There are approximately 5 000 000 people living in Chicago.
2. On average, there are two persons in each household in Chicago.
3. Roughly one household in twenty has a piano that is tuned regularly.
4. Pianos that are tuned regularly are tuned on average about once per year.
5. It takes a piano tuner about 2 h to tune a piano, including travel time.
6. Each piano tuner works 8 h in a day, 5 days in a week, and 50 weeks in a year.

From these assumptions we can compute:

- that the number of piano tunings in a single year in Chicago is: (5 000 000 persons in Chicago) / (2 persons/household) × (1 piano/20 households) × (1 piano tuning per piano per year) = 125 000 piano tunings per year in Chicago;
- that the average piano tuner performs (50 weeks/year) × (5 days/week) × (8 h/day)/(1 piano tuning per 2 h per piano tuner) = 1000 piano tunings per year per piano tuner;
- then dividing gives (125 000 piano tunings per year in Chicago)/(1000 piano tunings per year per piano tuner) = 125 piano tuners in Chicago, which is the answer that we sought.

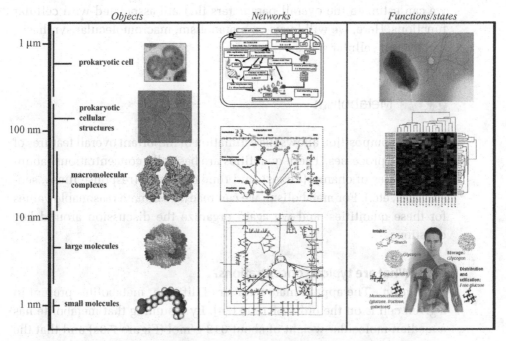

Figure 7.2 A multiscale view of metabolism, macromolecular synthesis, and cellular functions. Prokaryotic cell (*Synechocytis*) image from W. Vermaas, Arizona State University. Prokaryotic cell structures (purified carboxysomes) image from T. Yates, M. Yeager, and K. Dryden. Macromolecular complexes image ©2000, David S. Goodsell.

(2) *How far can a retrovirus diffuse before it falls apart?* A similar procedure that relies more on scientific principles can be used to answer this question. The half-life $t_{0.5}$ of retroviruses is measured to be about 5 to 6 h. The time constant for diffusion is

$$t_{\text{diff}} = l^2/D, \tag{7.1}$$

where l is the diffusion distance and D is the diffusion constant. Then, the distance $l_{0.5}$ that a virus can travel over a half life is

$$l_{0.5} = \sqrt{Dt_{0.5}}. \tag{7.2}$$

Using a numerical value for D of 6.5×10^{-8} cm^2/s that is computed from the Stokes–Einstein equation for a 100 nm particle (approximately the diameter of the retrovirus), the estimate is about 500 μm [25]. This is a fairly short distance, which limits how far a virus can go to infect a target cell.

A multiscale view The cellular composition of cells is complex. More complex yet is the intricate and coordinated web of complex functions that underlie the physiological state of a cell. We can view this as a multiscale relationship; see Figure 7.2. Based on cellular composition and other data, we can estimate the overall parameters that are associated with cellular functions. Here, we will focus on metabolism, macromolecular synthesis, and overall cellular states.

7.2   Metabolism

Biomass composition allows the estimation of important overall features of metabolic processes. These quantities are basically concentrations (abundance), rates of change (fluxes), and time constants (response times, sensitivities, etc.). For metabolism, we can readily estimate reasonable values for these quantities, and we again organize the discussion around key questions.

7.2.1 What are typical concentrations?

Estimation   The approximate number of different metabolites present in a given cell is on the order of 1000 [34]. By assuming that metabolite has a median molecular weight of about 312 g/mol (Figure 7.3a) and that the fraction of metabolites of the wet weight is 0.01, we can estimate a typical metabolite concentration of

$$x_{ave} \approx \frac{1 \ g/cm^3 \times 0.01}{1000 \times 312 \ g/mol} \approx 32 \ \mu M. \tag{7.3}$$

The volume of a bacterial cell is about 1 μm^3, or about 1 fL ($= 10^{-15}$ L). Since 1 μm^3 is a logical reference volume, we convert the concentration unit as follows:

$$1 \ \mu M = \frac{10^{-6} \ mol}{1 \ L} \times \frac{10^{-15} \ L}{1 \ \mu m^3} \times \frac{6 \times 10^{23} \ molecules}{mol} \tag{7.4}$$

$$= 600 \ molecules/\mu m^3 \tag{7.5}$$

This number is remarkably small. A typical metabolite concentration of 32 μM then translates into mere 19 000 molecules/μm^3. One would expect that such low concentrations would lead to slow reaction rates. However, metabolic reaction rates are fairly rapid. As discussed in Chapter 5, cells have evolved highly efficient enzymes to achieve high reaction rates that occur even in the presence of low metabolite concentrations.

Measurement   Experimentally determined ranges of metabolite concentrations fall around the estimated range; an example is provided in Table 7.2. Surprisingly, glutamate is at a concentration that falls within the 100 mM range in *E. coli*. Other important metabolites, such as ATP, tend to fall in the millimolar range. Intermediates of pathways are often

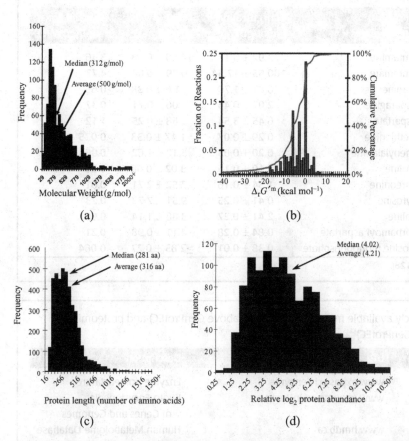

(a)   (b)

(c)   (d)

Figure 7.3 Details of *E. coli* K12 MG1655 composition and properties. (a) The average molecular weight of metabolites is 500 g/mol and the median is 312 g/mol. Molecular weight distribution. (b) Thermodynamic properties of the reactions in the iAF1260 reconstruction the metabolic network. (c) Size distribution of open reading frame (ORF) lengths or protein sizes: protein size distribution. The average protein length is 316 amino acids; the median is 281. Average molecular weight of *E. coli*'s proteins (monomers): 34.7 kDa, median: 30.828 kDa. (d) Distribution of protein concentrations: relative protein abundance distribution. See Refs [34] and [102] for details. Prepared by Vasiliy Portnoy.

in the micromolar range. Several on-line resources are now available for metabolic concentration data; see Table 7.3.

7.2.2 What are typical metabolic fluxes?

Rates of diffusion In estimating reaction rates we first need to know if they are diffusion limited. Typical cellular dimensions are on the order of micrometers, or less. The diffusion constants for metabolites are on the order of 10^{-5} cm^2/s and 10^{-6} cm^2/s. These figures translate into diffusional response times that are on the order of

$$t_{\mathrm{diff}} = \frac{l^2}{D} = \frac{(10^{-4} \ \mathrm{cm})^2}{10^{-5} \ \mathrm{to} \ 10^{-6} \ \mathrm{cm}^2/\mathrm{s}} \approx 1\text{–}10 \ \mathrm{ms} \qquad (7.6)$$

Table 7.2 Measured and predicted parameters for *E. coli* growing on minimal media

	Abbr.	Intermediates	Concentration (μmol/gDW)	PERC (1/min)	Measured flux (mmol/(gDW h))
1	Gln	Glutamine	3.92 ± 0.17	14.29 ± 6.28	3.36
2	Glu	Glutamate	100.55 ± 17.54	0.79 ± 0.04	4.77
4	Ala	Alanine	6.81 ± 1.70	≥1.6 ± 0.38	≥0.65
5	Asn	Asparagine	2.02 ± 0.46	3.06 ± 0.71	0.37
6	Asp	Aspartate	6.45 ± 3.54	2.88 ± 0.25	1.12
7	Met	Methionine	0.29 ± 0.07	≥1.47 ± 0.33	≥0.025
8	Phe	Phenylalanine	0.20 ± 0.03	5.12 ± 1.62	0.063
9	Pro	Proline	1.10 ± 0.15	≥3.02 ± 0.66	≥0.20
10	Thr	Threonine	1.34 ± 0.16	7.52 ± 2.71	0.61
11	Tyr	Tyrosine	0.41 ± 0.25	9.51 ± 7.73	0.23
12	Val	Valine	2.41 ± 0.27	3.88 ± 1.14	0.56
13		Carbamoyl aspartate	0.84 ± 0.28	4.11 ± 0.98	0.21
14	IMP	Inosine monophosphate	0.38 ± 0.01	≥2.83 ± 0.27	≥0.064

Taken from Ref. [128].

Table 7.3 Publicly available metabolic resources (above GenProtEC) and proteomic resources (from GenProtEC)

Database	URL	Description
ENZYME	www.expasy.ch/enzyme	Enzyme nomenclature
KEGG	www.genome.ad.jp/kegg	Kyoto Encyclopedia of Genes and Genomes
HMDB	www.hmdb.ca	Human Metabolome Database
BioCyc	www.biocyc.org/ecocyc	Pathway and Genome Database
LIGAND	www.genome.ad.jp/dbget/ligand.html	Biochemical Reactions Database
HML	www.metabolibrary.ca	Human Metabolome Library
MMCD	mmcd.nmrfam.wisc.edu	Madison Metabolomics Consortium Database
BRENDA	www.brenda-enzymes.info/	The Comprehensive Enzyme Information System
SABIO-RK	sabio.villa-bosch.de	Reaction Kinetics Database
GenProtEC	genprotec.mbl.edu	*E. coli* Genome and Proteome Database
PRIDE	www.ebi.ac.uk/pride	The PRIDE Proteomics IDEntifications Database
ExPASy	br.expasy.org	Expert Protein Analysis System
MCBI	www.ctaalliance.org/MCBI/Proteomics.html	Michigan Proteomics Consortium
SWISS-PROT	ca.expasy.org/sprot/	Protein Knowledge Database
OPD	bioinformatics.icmb.utexas.edu/OPD	Open Proteomics Database
Proteomics World	www.proteomicworld.org/DatabasePage.html	Resources for Proteomics and Protein Expression

Assembled by Vasiliy Portnoy.

or faster. The metabolic dynamics of interest are much slower than milliseconds. Although more detail about the cell's finer spatial structure is becoming increasingly available, it is unlikely, from a dynamic modeling standpoint, that spatial concentration gradients will be a key concern for dynamic modeling of metabolic states in bacterial cells [124].

Estimating maximal reaction rates Reaction rates in cells are limited by the achievable kinetics. Few collections of enzyme kinetic parameters are available in the literature; see Table 7.3. One observation from such collections is that the bimolecular association rate constant k_1 for a substrate (S) to an enzyme (E),

$$S + E \xrightarrow{k_1},$$ (7.7)

is on the order of 10^8 M^{-1} s^{-1}. This numerical value corresponds to the estimated theoretical limit, due to diffusional constraints [39]. The corresponding number for macromolecules is about three orders of magnitude lower.

Using the order-of-magnitude values for concentrations of metabolites given above and for enzymes in the next section, we find the representative association rate of substrate to enzymes to be on the order of

$$k_1 se = 10^8 \ (M \times s)^{-1} \times 10^{-4} \ M \times 10^{-6} \ M = 0.01 \ M/s,$$ (7.8)

which translates into about

$$k_1 se = 10^6 \ \text{molecules}/(\mu m^3 \ s),$$ (7.9)

that is, only 1 million molecules per cubic micrometer per second. However, the binding of the substrate to the enzyme is typically reversible, and a better order-of-magnitude estimate for *net* reaction rates is obtained by considering the release rate of the product from the substrate–enzyme complex, X. This release step tends to be the slowest step in enzyme catalysis [10, 11, 26]. Typical values for the release rate constant k_2,

$$X \xrightarrow{k_2} P + E,$$ (7.10)

are 100–1000 s^{-1}. If the concentration of the intermediate substrate–enzyme complex X is on the order of 1 μM, we get a release rate of about

$$k_2 x = 10^4 \ \text{to} \ 10^5 \ \text{molecules}/(\mu m^3 \ s).$$ (7.11)

We can compare the estimate in Eq. (7.9) with observed metabolic fluxes; see Table 7.4. Uptake and secretion rates of major metabolites during bacterial growth represent high flux pathways.

Table 7.4 Typical metabolic fluxes measured in *E. coli* K12 MG1655 grown under oxic and anoxic conditions

	Rate (mmol/(gDW h))		Molecules/(μm^3 s)	
Flux name	Anoxic	Oxic	Anoxic	Oxic
Glucose uptake	17.9 ± 1.2	9.02 ± 0.23	8.35×10^5	4.21×10^5
Oxygen uptake	0	14.92 ± 0.21	0	6.96×10^5
Formate secretion	15.8 ± 1.8	3.51 ± 0.47	7.37×10^5	1.57×10^5
Acetate secretion	10.9 ± 0.8	3.37 ± 0.02	5.13×10^5	1.64×10^5
Ethanol secretion	7.4 ± 0.6	0	3.50×10^5	0
Succinate secretion	1.1 ± 0.4	0	5.13×10^5	0
Lactate secretion	0.2 ± 0.1	0	9.33×10^5	0

Measured kinetic constants   There are now several accessible sources of information that contain kinetic data for enzymes and the chemical transformation that they catalyze. For kinetic information, both BRENDA and SABIO-RK [125] are resources of literature curated constants, including rates and saturation levels; Table 7.3. Unlike stoichiometric information, which is universal, kinetic parameters are highly condition dependent. *In vitro* kinetic assays typically do not represent *in vivo* conditions. Factors such as cofactor binding, pH, and unknown interactions with metabolites and proteins are likely causes.

Thermodynamics   While computational prediction of enzyme kinetic rates is difficult, obtaining thermodynamic values is more feasible. Estimates of metabolite standard transformed Gibbs energy of formation can be derived using an approach called the *group contribution method* [75]. This method considers a single compound as being made up of smaller structural subgroups. The metabolite standard Gibbs energy of formation associated with structural subgroups commonly found in metabolites is available in the literature and in the NIST database [2]. To estimate the metabolite standard Gibbs energy of formation of the entire compound, the contributions from each of the subgroups are summed along with an origin term. The group contribution approach has been used to estimate standard transformed Gibbs energy of formation for 84% of the metabolites in the genome-scale model of *E. coli* [34, 44, 45].

Thermodynamic values can also be obtained by integrating experimentally measured parameters and algorithms which implement sophisticated theory from biophysical chemistry [8, 9]. Combining this information with the results from the group contribution method provides standard

Table 7.5 Turnover times for the glycolytic intermediates in the red blood cell. The glycolytic flux is assumed to be 1.25 mM/h = 0.35 μM/s and the Rapoport–Luebering shunt flux is about 0.5 mM/h

Compound	Concentration (mM)	Turnover time (s)
G6P	0.038 ± 0.012	109
F6P	0.016 ± 0.03	46
FDP	0.0076 ± 0.004	21
DHAP	0.140 ± 0.08	420
GA3P	0.0067 ± 0.001	19
1,3DPG	0.0004	1.1
3PG	0.045	120
2PG	0.014 ± 0.005	40
PEP	0.017 ± 0.002	50
PYR	0.077 ± 0.05	240
LAC	1.10 ± 0.50	3120

Table adapted from Ref. [62].

transformed Gibbs energy of formation for 96% of the reactions in the genome-scale *E. coli* model [34]; see Figure 7.3b.

7.2.3 What are typical turnover times?

As outlined in Chapter 2, turnover times can be estimated by taking that ratio of the concentration relative to the flux of degradation. Both concentrations and fluxes have been estimated above. Some specific examples of estimated turnover times are now provided.

Glucose turnover in rapidly growing E. coli cells   With an intracellular concentration of glucose of 1 to 5 mM, the estimate of the internal glucose turnover time is

$$\tau_{glu} = \frac{(6-30) \times 10^5 \text{ molecules/cell}}{4.2 \times 10^5 \text{ to } 8.4 \times 10^5 \text{ molecules/}(\mu m^3 \text{ s})} = 1 \text{ to } 8 \text{ s.} \quad (7.12)$$

Response of red-cell glycolytic intermediates   A typical glycolytic flux in the red cell is about 1.25 mM/h. By using this number and measured concentrations, we can estimate the turnover times for the intermediates of glycolysis by simply using

$$t_R = \frac{x_{avg}}{1.25 \text{ mM/h}}. \quad (7.13)$$

The results are shown in Table 7.5. We see the sharp distribution of turnover times that appears. Note that the turnover times are set by the

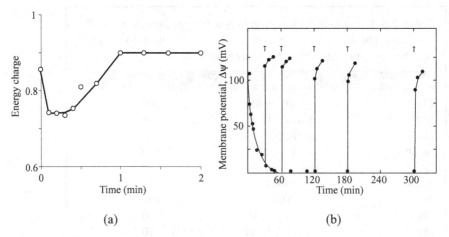

Figure 7.4 Responses in energy transduction processes in cells. (a) Effect of addition of glucose on the energy charge of Ehrlich ascites tumor cells. Redrawn based on Ref. [14]. (b) Transient response of the transmembrane gradient, from Ref. [67]. Generation of a protonmotive force in energy-starved *Streptococcus cremoris* upon addition of lactose (indicated by arrows) at different times after the start of starvation ($t = 0$).

relative concentrations, since the flux through the pathway is the same. Thus, the least abundant metabolites will have the fastest turnover. At a constant flux, the relative concentrations are set by the kinetic constants.

Response of the energy charge   Exchange of high-energy bonds between the various carriers is on the order of minutes. The dynamics of this energy pool occur on the middle time scale of minutes, as described earlier; see Figure 7.4.

Response of transmembrane charge gradients   Cells store energy by extruding protons across membranes. The consequence is the formation of an osmotic and charge gradient that results in the so-called *protonmotive force*, denoted as $\Delta\mu_{H^+}$. It is defined by

$$\Delta\mu_{H^+} = \Delta\Psi - Z\Delta pH, \tag{7.14}$$

where $\Delta\Psi$ is the charge gradient and ΔpH is the hydrogen ion gradient. The parameter Z takes a value of about 60 mV under physiological conditions. The transient response of gradient establishment is very rapid; see Figure 7.4.

Conversion between different forms of energy   If energy is to be readily exchanged between transmembrane gradients and the high-energy phosphate bond system, their displacement from equilibrium should be about the same. Based on typical ATP, ADP, and P_i concentrations, one can

Table 7.6 Typical values (mV) for the transmembrane electrochemical potential gradient

Material	Conditions	$\Delta\psi$	$-60\,\Delta pH$	$\Delta\tilde{\mu}_{H^+}$	Ref.
Liver mitochondria	State 4	168	48	216	[81]
Brown fat mitochondria	Proton channel open	79	−25	54	[85]
	Proton channel inhibited	134	−95	229	[85]
Heart sub-mitochondrial particles	NADH substrate	145	0	145	[114]
E. coli cells	Respiring	100	−105	205	[130]
Chloroplasts	Light	0	180	180	[105]
Chromatophores	Dark	12	65	78	[111]
	Light	89	106	195	[111]

Reproduced from Ref. [67].

calculate the transmembrane gradient to be about −180 mV. Table 7.6 shows that observed values for the transmembrane gradient $\Delta\tilde{\mu}$ are on the order of −180 to −220 mV. It is interesting to note that the maximum gradient that a bilipid layer can withstand is on the order of −280 mV [67], based on electrostatic considerations.

7.2.4 What are typical power densities?

The metabolic rates estimated above come with energy transmission through key cofactors. The highest energy production and dissipation rates are associated with energy transducing membranes. The ATP molecule is considered to be an energy currency of the cell, allowing one to estimate the power density in the cell/organelle based on the ATP production rate.

Power density in mitochondria The rate of ATP production in mitochondria can be measured. Since we know the energy in each phosphate bond and the volume of the mitochondria, we can estimate the volumetric rate of energy production in the mitochondria.

Reported rates of ATP production in rat mitochondria from succinate are on the order of 6×10^{-19} mol ATP per mitochondria per second [63], taking place in a volume of about 0.27 μm^3. The energy in the phosphate bond is about −52 kJ/mol ATP at physiological conditions. These numbers lead to the computation of a per unit volume energy production rate of 2.2×10^{-18} mol ATP/(μm^3 s), or 10^{-13} W/μm^3 (0.1 pW/μm^3).

Power density in chloroplast of green algae In *Chlamydomonas reinhardtii*, the rate of ATP production of chloroplast varies between 9.0×10^{-17} and 1.4×10^{-16} mol ATP per chloroplast per second depending on the light intensity [16, 23, 79, 104] in the volume of 17.4 μm^3 [40]. Thus, the volumetric energy production rate of chloroplast is on the order of 5×10^{-18} mol ATP/(μm^3 s), or 3×10^{-13}W/μm^3 (0.3 pW/μm^3).

Power density in rapidly growing E. coli cells: A similar estimate of energy production rates can be performed for microorganisms. The aerobic glucose consumption of *E. coli* is about 10 mmol/(gDW h). The weight of a cell is about 2.8×10^{-13} gDW. The ATP yield on glucose is about 17.5 ATP/glucose. These numbers allow us to compute the energy generation density from the ATP production rate of 1.4×10^{-17} mol ATP per cell per second. These numbers lead to the computation of the power density of 7.3×10^{-13} W/μm^3 O_2 (0.7 pW/μm^3), which is similar to the numbers computed for the mitochondria and chloroplast above.

7.3   Macromolecules

We now look at the abundance, concentration, and turnover rates of macromolecules in the bacterial cell. We are interested in the genome, RNA, and protein molecules.

7.3.1 What are typical characteristics of a genome?

Sizes of genomes vary significantly amongst different organisms; see Table 7.7. For bacteria, they vary from about 0.5 to 9 million base pairs. The key features of the *E. coli* K-12 MG1655 genome are summarized in Table 7.8. There are about 4500 ORFs on the genome of an average length of about 1 kb. This means that the average protein size is 316 amino acids; see Figure 7.3c.

The number of RNA polymerase binding sites is estimated to be about 2800, leading to a estimate of about 1.6 ORFs ($\approx 4.4 \times 10^6/2400$) per transcription unit. There are roughly 3000 copies of the RNA polymerase present in an *E. coli* cell [122]. Thus, if 1000 of the transcription units are active at any given time, there are only two or three RNA polymerase molecules available for each transcription unit. The promoters have different binding strengths and thus recruit a different number of RNA polymerase molecules each. ChIP-chip data can be used to estimate this distribution.

7.3.2 What are typical protein concentrations?

Cells represent a fairly dense solution of protein. One can estimate the concentration ranges for individual enzymes in cells. If we assume that the cell has about 1000 proteins with an average molecular weight of 34.7 kDa, as is typical for an *E. coli* cell, see Figure 7.3c, and given the fact

Table 7.7 Genome sizes. A selection of representative genome sizes from the rapidly growing list of organisms whose genomes have been sequenced

Organism	Genome size (bp)	Genes (#)
Organelle/viruses		
Phi-X 174	5 386	10
Human mitochondrion	16 569	37
Epstein–Barr virus (EBV)	172 282	80
Mimivirus	1 181 404	1 262
Archaea		
Nanoarchaeum equitans	490 885	552
Methanococcus jannaschii	1 664 970	1 783
Methanobacterium		
thermoautotrophicum	1 751 377	2 008
Cyanidioschyzon merolae	16 520 305	5 331
Prokaryotes		
Chlamydia trachomatis	1 042 519	936
Helicobacter pylori	1 667 867	1 589
Haemophilus influenzae	1 830 138	1 738
Thermotoga maritima	1 860 725	1 879
Streptococcus pneumoniae	2 160 837	2 236
Neisseria meningitidis	2 184 406	2 185
Vibrio cholerae	4 033 460	3 890
Bacillus subtilis	4 214 814	4 779
Escherichia coli	4 639 221	4 288
Eukaryotes		
Schizosaccharomyces pombe	12 462 637	4 929
Saccharomyces cerevisiae	12 495 682	5 770
Caenorhabditis elegans	100 258 171	19 427
Drosophila melanogaster	122 653 977	13 379
Anopheles gambiae	278 244 063	13 683
Plants		
Arabidopsis thaliana	115 409 949	~28 000

Adapted from Ref. [3].

that the cellular biomass is about 15% protein, we get

$$e_{\text{tot}} \approx \frac{1 \text{ g/cm}^3 \times 0.15}{1000 \times 34\,700 \text{ g/mol}} = 4.32 \text{ μM.} \tag{7.15}$$

This estimate is, indeed, the region into which the *in vivo* concentration of most proteins falls. It corresponds to about 2500 molecules of a

Table 7.8 Some features of the E. coli genome

Process	Size
Genome size (bp)	4 639 221
Number of ORFs (genes)	4 505
Average ORF size (bp)	≈1 000
RNAP binding sites	≈2 800
Protein coding genes (%)	87.80
Stable RNA genes (%)	0.80
rRNA (genes)	7
tRNA (genes)	86
Noncoding repeats (%)	0.70
Regulatory and other functions (%)	11

From Ref. [8].

Table 7.9 Rate characteristics of macromolecular synthesis and degradation. DNA is stable, and so are tRNA and rRNA molecules. About 2–20% of the protein is turned over per hour

Process	Rate
DNA replication	900 bp/s per replication fork
RNA polymerase	40–50 bp/s
Ribosome	12–21 peptide bonds per second
mRNA half life	4–6 min

particular protein molecule per cubic micrometer. As with metabolites, there is a significant distribution around the estimate of Eq. (7.15). Important proteins, such as the enzymes catalyzing major catabolic reactions, tend to be present in higher concentrations, and pathways with smaller fluxes have their enzymes in lower concentrations. It should be noted that we are not assuming that all these proteins are in solution; the above number should be viewed more as a molar density.

The distribution about this mean is significant; Figure 7.3d. Many of the proteins in E. coli are in concentrations as low as a few dozen per cell. The E. coli cell is believed to have about 200 major proteins, which brings our estimate for the abundant ones to about 12 000 copies per cell. The E. coli cell has a capacity to carry about 2.0–2.3 million protein molecules.

7.3.3 What are typical fluxes?

The rates of synthesis of the major classes of macromolecules in E. coli are summarized in Table 7.9. The genome can be replicated in 40 min with two replication forks. This means that the speed of DNA polymerase is estimated to be

$$\text{rate of DNA polymerase} = \frac{4.4 \times 10^6 \text{ bp}}{2 \times 40 \times 60} \approx 900 \text{ bp/s per fork.} \quad (7.16)$$

RNA polymerase is much slower, at 40–50 bp/s, and the ribosomes operate at about 12–21 peptide bonds per ribosome per second.

Protein synthesis capacity in E. coli We can estimate the number of peptide bonds (pb) produced by *E. coli* per second. To do so we will need the rate of peptide bond formation by the ribosome (12 to 21 bp per ribosome per second [21, 27, 127]) and number of ribosomes present in the *E. coli* cell (7×10^3 to 7×10^4 ribosomes/cell, depending on the growth rate [21]). So, the total number of peptide bonds that *E. coli* can make per second is on the order of 8×10^4 to 1.5×10^6 pb/cell.

The average size of a protein in *E. coli* is about 320 amino acids. At about 45 to 60 min doubling time, the total amount of protein produced by *E. coli* per second is ~300 to 900 protein molecules/cell. This is equivalent to $(1 \text{ to } 3) \times 10^6$ molecules per cell per hour as a function of growth rate, which is about the total number of proteins per cell given above.

Maximum protein production rate from a single gene in murine cells The total amount of the protein formed from a single gene in the mammalian cell can be estimated based on the total amount of mRNA present in the cytoplasm from a single gene, the rate of translation of the mRNA molecule by ribosomes, and the ribosomal spacing [108]. Additional factors needed include gene dosage (here taken as unity), rate of mRNA degradation, velocity of the RNA polymerase II molecule, and the growth rate [108].

Murine hybridoma cell lines are commonly used for antibody production. For this cell type, the total amount of mRNA from a single antibody-encoding gene in the cytoplasm is of the order of 40 000 mRNA molecules/cell [37, 109], the ribosomal synthesis rate is on the order of 20 nucleotides/s [98], and the ribosomal spacing on the mRNA is between 9 and 100 nucleotides [24, 98]. Multiplying these numbers, we can estimate the protein production rate in a hybridoma cell line to be approximately 3000 to 6000 protein molecules per cell per second.

7.3.4 What are typical turnover times?

The assembly of new macromolecules such as RNA and proteins requires the source of nucleotides and amino acids. These building blocks are generated by the degradation of existing RNA molecules and proteins. Many cellular components are constantly degraded and synthesized. This process is commonly characterized by the turnover rates and half-lives. Intracellular protein turnover is experimentally assessed by an addition of an isotope-labeled amino acid mixture to the normally growing or nongrowing cells [71, 99]. It has been shown that the rate of breakdown of an individual protein is on the order of 2–20% per hour in *E. coli* culture [71, 84].

7.4   Cell growth and phenotypic functions

7.4.1 What are typical cell-specific production rates?

The estimated rates of metabolism and macromolecular synthesis can be used to compute various cellular functions and their limitations. Such computations can in turn be used for bioengineering design purposes, environmental roles and impacts of microorganisms, and for other purposes. We provide a couple of simple examples.

Limits on volumetric productivity   *E. coli* is one of the most commonly used host organisms for metabolic engineering and overproduction of metabolites. In many cases, the glycolytic flux acts as a carbon entry point to the pathway for metabolite overproduction [31, 56, 126, 129]. Thus, the substrate uptake rate (SUR) is one of the critical characteristics of the productive capabilities of the engineered cell.

Let us examine the wild-type *E. coli* grown on glucose under anoxic conditions. As shown in Table 7.4, the SUR is on the order of 15–20 mmol glucose per gDW per hour, which translates into 1.5 g glucose per liter per hour at cell densities of 2–4 gDW/L. Theoretically, if all the carbon source (glucose) is converted to the desired metabolite, the volumetric productivity will be approximately 3 g/(L h).

The amount of cells present in the culture plays a significant role in production potential. In industrial settings, the cell density is usually higher, which increases the volumetric productivity. Some metabolic-engineered-strain designs demonstrate a higher SUR [97], which also leads to the increase in volumetric productivity.

Photoautotrophic growth   *Chlorella vulgaris* is a single-celled green alga that uses light to generate energy necessary for growth. At the top rate of photosynthesis, the specific oxygen production rate (SOPR) can be estimated to be between 20–400 fmol O_2 per cell per hour [69]. Algae biotechnology is drawing increasing interest due to its potential for production of biofuels and fine chemicals [69]. However, a lack of suitable photobioreactors (PBRs) makes the cost of algally derived compounds high. One of the

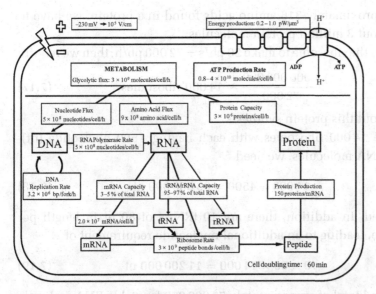

Figure 7.5 Order-of-magnitude calculations applied to bacterial metabolism. Rates such as metabolic flux, mRNA and DNA synthesis, protein synthesis, etc. were estimated using the order-of-magnitude approach presented in this chapter. Rates were calculated for *E. coli* grown on glucose minimal media doubling every 60 min. All rates, except for energy production, are presented per one doubling time (1 h).

key limiting factors for PBRs is the light source; however, light-emitting diodes (LEDs) can be employed for these purposes.

Let us now use order-of-magnitude calculations to estimate the light requirement for an algae PBR using *C. vulgaris* as a model organism. Given the fact that maximum photosynthetic efficiency of *C. vulgaris* is below 50% [66, 83, 96] and 1 mol of photons (680 nm) is equivalent to 50 W, we can estimate that in order to sustain the SOPR of 100 fmol O_2 per cell per hour each cell must receive 40 pW equivalent of photons [69]. A conventional LED can provide 0.3 mW/cm^2 or 0.1 mW per LED. With a cell density close to 10^9 cells/ml and 80 cm^3 volume of the reactor [69], the PBR must include close to a 100 LEDs to sustain the growth of algae and oxygen production.

7.4.2 Balancing the fluxes and composition in an entire cell

The approximate calculation procedures presented can be used to estimate the overall flows of mass and energy in a bacterial cell. Proteins are 55% of the dry weight of cells and their most energetically costly component, so let us begin such a computation with the assumption that there are about 10^9 amino acids found in the proteins of a single cell. With this starting point and the various data given in this chapter, we can roughly estimate all the major flows using a 60 min doubling time (see Figure 7.5):

- With approximately 316 amino acids found in a protein, we have to make about 3 million protein molecules.
- If we take the ribosome to make 20 pb/s = 72 000 pb/h, then we require

$$\frac{1\,000\,000\,000}{72\,000} \approx 14\,000 \text{ ribosomes} \tag{7.17}$$

to carry out this protein synthesis.

- To make 14 000 ribosomes with each having 4 500 nucleotide (nt)-length RNA molecules, we need

$$14\,000 \times 4500 = 63\,000\,000 \text{ nt} \tag{7.18}$$

assembled. In addition, there are 10 tRNAs of 80 nt in length per ribosome, leading to an additional nucleotide requirement of

$$10 \times 80 \times 14\,000 = 11\,200\,000 \text{ nt} \tag{7.19}$$

for a grand total of approximately 75 000 000 for stable RNA molecule synthesis.

- The total nucleotide synthesis for RNA will be 3000 RNA polymerase molecules synthesizing at the rate of 50 nt/s or

$$3000 \times 50 \times 3600 = 540\,000\,000 \text{ nt/h.} \tag{7.20}$$

- The fraction of RNA that is mRNA is 0.03 to 0.05 [103], or

$$540\,000\,000 \times (0.03 \text{ to } 0.05) \approx (16 \text{ to } 25.0) \times 10^6 \text{ nt per cell per hour.} \tag{7.21}$$

If the average mRNA length is 1100 nt, then the cell needs to make, on average, 20 000 transcripts in 1 h.

- We have to make 3 000 000 proteins from 20 000 transcripts, or about 150 protein molecules per transcript.
- The transcripts have a finite half-life. On average, each transcript has a 5 min lifetime, or 300 s. Owing to structural constraints, a ribosome can only bind every 50 nt to the mRNA, producing a maximum ribosomal loading of about 20 ribosomes per transcript. The rate of translation is 20 pb/s. With the average length of the peptide being 316 amino acids, we can produce 1.25 proteins/s. This calculation estimated the maximum protein production from one transcript on the order of 375 protein molecules per transcript.
- To synthesize the genome, we need $2 \times 4\,500\,000 = 9\,000\,000$ nt to make the double-stranded DNA.
- Thus, the total metabolic requirement of amino acids and nucleotides in *E. coli* per doubling is 1×10^9 amino acids per cell per hour and 5×10^8 nt per cell per hour.

These are the approximate overall material requirements. We also need energy to drive the process. Using Table 7.4, we can estimate energy requirements for *E. coli* under oxic and anoxic conditions.

- Aerobically, at a doubling time of 1 h, the glucose uptake rate is about 10 mmol/(gDW h), which is equivalent to 1.5×10^9 molecules of glucose per cell per doubling. At 17.5 ATP produced per glucose, the corresponding energy production is 3×10^{10} ATP per cell per doubling.
- Anaerobically, at a doubling time of 1.5 h, the glucose uptake rate is about 18 mmol/(gDW h), which is equivalent to 4.5×10^9 molecules of glucose per cell per doubling. At 3 ATP per glucose the corresponding energy production is 1.4×10^{10} molecules ATP per cell per doubling.

7.5 Summary

➤ Data on cellular composition and overall rates are available.

➤ Order-of-magnitude estimation procedures exist through which one can obtain the approximate values for key quantities.

➤ In this fashion, typical concentrations, fluxes, and turnover times can be estimated.

➤ An approximate quantitative overall multiscale framework can be obtained for the function of complex biological processes.

## CHAPTER 8

Stoichiometric structure

Part I of this book introduced the basics of dynamic simulation. The process for setting up dynamic equations, their simulation, and processing of the output was presented in Chapter 3. Several concepts of dynamic analysis of networks were illustrated through the use of simple examples of chemical reaction mechanisms in Chapters 4 through 6. Most of these examples were conceptual and had limited direct biological relevance. In Chapter 7 we began to estimate the numerical values and ranges for key quantities in dynamic models. With this background, we now begin the process of addressing issues that are important when one builds realistic dynamic models of biological functions. We start by exploring the consequences of reaction bilinearity and that of the stoichiometric structure of a network. In Part II we extend the material in this chapter to well-known metabolic pathways.

8.1   Bilinear biochemical reactions

Bilinear reactions

$$x + y \rightarrow z \tag{8.1}$$

characterize biochemical reaction networks. Two molecules come together to form a new molecule through the breaking and forming of covalent bonds, or a complex through the formation of hydrogen bonds. As illustrated with the pool formations in the bilinear examples in Chapter 4, such reactions come with moiety exchanges.

Enzyme classification Enzyme-catalyzed reactions are classified into seven categories by Enzyme Commission (EC) numbers; see Figure 8.1a.

(a)

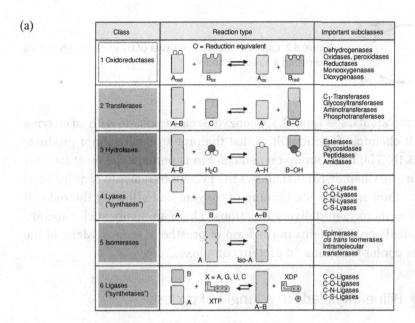

(b)

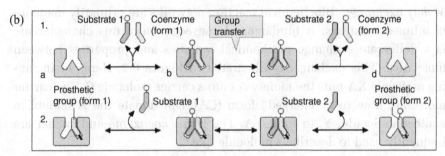

Figure 8.1 The bilinear nature of biochemical reactions. (a) The classification of enzyme-catalyzed reactions into seven categories by the Enzyme Commission (EC) number system. (b) The detailed view of the role of cofactors and prosthetic groups in enzyme-catalyzed reactions. Both images from Ref. [68] (reprinted with permission).

These categories are: oxidoreductases, transferases, hydrolases, lyases, isomerases, and ligases. All these chemical transformations are bilinear with the exception of isomerases, which simply rearrange a molecule without the participation of other reactants. Thus, the vast majority of biochemical reactions are bilinear. An overall pseudo-elementary representation (i.e., without treating the enzyme itself as a reactant and just representing the uncatalyzed reaction) is bilinear.

Cofactors and coenzymes There are certain cofactors that are involved in many biochemical reactions. These cofactors are involved in group transfer reactions; Figure 8.1b. They can transfer various chemical moieties or

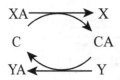

Figure 8.2 Carrier (C) mediated transfer of chemical moiety A from compound X to compound Y.

redox equivalents; see Table 8.1. Coenzymes can associate with an enzyme to give it chemical functionalities that the protein itself cannot produce; Figure 8.1b. The heme group on hemoglobin is perhaps the most familiar example (see Chapter 13) that allows the protein tetramer to acquire a ferrous ion, thus enabling the binding of oxygen, and allowing the red cell to perform its oxygen delivery functions. There are many such capabilities "grafted" onto proteins in the form of prosthetic groups. Many of the vitamins confer functions on protein complexes.

8.2   Bilinearity leads to a tangle of cycles

Moiety exchange   Biochemical reaction networks are primarily made up of bilinear reactions. A fundamental consequence of this characteristic is a deliberate exchange of chemical moieties and properties between molecules. This exchange is illustrated in Figure 8.2. Here, an incoming molecule XA puts the moiety A onto a carrier molecule C. The carrier molecule, now in a "charged" form (CA), can donate the A moiety to another molecule, Y, to form YA. The terms *coenzyme* and *carrier* are frequently used to describe a molecule like C.

Formation of cycles   The ability of bilinear reactions to exchange moieties in this fashion leads to the formation of distribution networks of chemical moieties and other properties of interest through the formation of a deliberate "supply-chain" network. The structure of such a network must be thermodynamically feasible and conform to environmental constraints.

Bilinearization in biochemical reaction networks leads to a "tangle of cycles," where different moieties and properties are being moved around the network. While a property of all biochemical networks, this trafficking of chemical and other properties is best known in metabolism. The major chemical properties that are being exchanged in metabolism are summarized in Table 8.1. These properties include energy, redox potential, one-carbon units, two-carbon units, amide groups, amine groups, etc. We now consider some specific cases.

Example: Redox and energy trafficking in the core E. coli metabolic pathways Energy metabolism revolves around the generation of redox potential

Table 8.1 Some activated carriers or coenzymes in metabolism

Coenzyme/carrier	Examples of chemical groups transferred	Dietary precursor in mammals
Carbon		
Biocytin	CO_2	Biotin
Tetrahydrofolate	One-carbon groups	Folate (B9)
Coenzyme A (CoA)	Acyl groups	Pantothenic acid (B5) and other compounds
Thiamine pyrophosphate (TPP)	Aldehydes	Thiamine (B1)
S-Adenosyl methionine	Methyl groups	
Cyano/5′-deoxyadenosyl cobalamin	Rearrangement of vicinal −H and −R groups	Coenzyme B12
Energy		
ATP (NTPs)	P_i or PP_i	
Redox		
Flavin adenin dinucleotide	Electrons	Riboflavin (B2)
Lipoate	Electrons and acyl groups	
Nicotinamide adenine dinucleotide (NAD or NADP)	Hydride ion	
Coenzyme Q	Electrons and protons	Vitamin Q
Cytochromes/heme	Electrons	Iron
Nitrogen		
Pyridoxal phosphate	Amino groups	Pyridoxine (B6)
Glutamate (N_2-fixing plants)	Amino groups	

Modified from Ref. [47].

and chemical energy in the form of high-energy phosphate bonds. The degradation of substrates through a series of chemical reactions culminates in the storage of these properties on key carrier molecules; see Table 8.1.

The core metabolic pathways in *E. coli* illustrate this feature. The transmission of redox equivalents through this core set of pathways is shown in Figure 8.3a. Each pathway is coupled to a redox carrier in a particular way. This pathway map can be drawn to show the cofactors rather than the primary metabolites and the main pathways (Figure 8.3b). This figure clearly shows how the cofactors interact and how the bilinear property of the stoichiometry of the core set of pathways leads to a tangle of cycles among the redox carriers.

Example: Protein trafficking in signaling pathways Although the considerations above are illustrated using well-known metabolic pathways, these

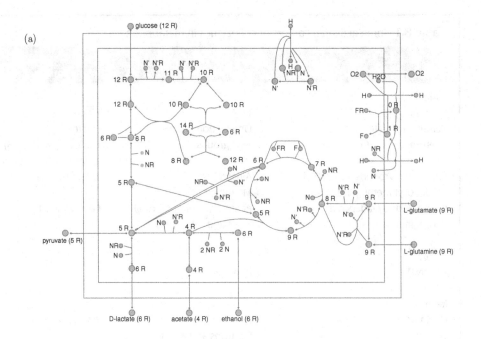

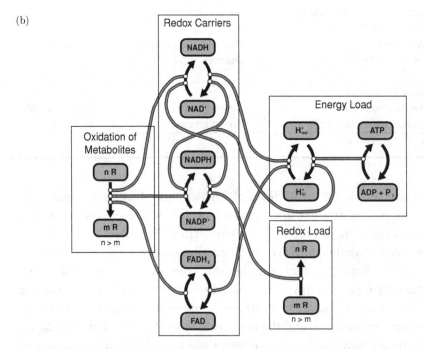

Figure 8.3 The tangle of cycles in trafficking of redox potential (R) in *E. coli* core metabolic pathways. (a) A map organized around the core pathways. (b) The tangle of cycles seen by viewing the cofactors and how they are coupled. Prepared by Jeff Orth.

same features are also being observed in signaling pathways. Incoming molecules (ligands) trigger a well-defined series of charging and discharging of the molecules that make up a signaling network, most often with a phosphate group.

8.3   Trafficking of high-energy phosphate bonds

Given the bilinear nature of biochemical reaction networks and the key role that cofactors play, we begin the process of building biologically meaningful simulation models by studying the use and formation of high-energy phosphate bonds. Cellular energy is stored in high-energy phosphate bonds in ATP. The dynamic balance of the rates of use and formation of ATP is thus a common denominator in all cellular processes, and thus foundational to the living process. We study the dynamic properties of this system in a bottom-up fashion.

8.3.1 The basic structure of the "core" module

The mass balances   The basic structure of the trafficking of high-energy phosphate bonds on the adenosine carrier is given by the following kinetic equations (see Figure 8.4a):

$$\frac{d\,\text{ATP}}{dt} = -v_{\text{use}} + v_{\text{form}} + v_{\text{distr}}, \tag{8.2}$$

$$\frac{d\,\text{ADP}}{dt} = v_{\text{use}} - v_{\text{form}} - 2v_{\text{distr}}, \tag{8.3}$$

$$\frac{d\,\text{AMP}}{dt} = v_{\text{distr}}, \tag{8.4}$$

where v_{use} is the rate of use of ATP, v_{form} is the rate of formation of ATP, and v_{distr} is the redistribution of the phosphate group among the adenosine phosphates by adenylate kinase.

The reaction rates   The mass action forms of these basic reaction rates are

$$\begin{aligned}
v_{\text{use}} &= k_{\text{use}}\text{ATP}, \quad v_{\text{form}} = k_{\text{form}}\text{ADP},\\
v_{+\text{distr}} &= k_{\text{distr}}\text{ADP}^2, \quad v_{-\text{distr}} = k_{-\text{distr}}\text{ATP} \times \text{AMP}.
\end{aligned} \tag{8.5}$$

Numerical values   The approximate numerical values of the parameters in this system can be estimated. The ATP concentration is about 1.6 mM, the ADP concentration is about 0.4 mM, and the AMP concentration is about 0.1 mM. The total adenosine phosphates are thus about 2.1 mM.

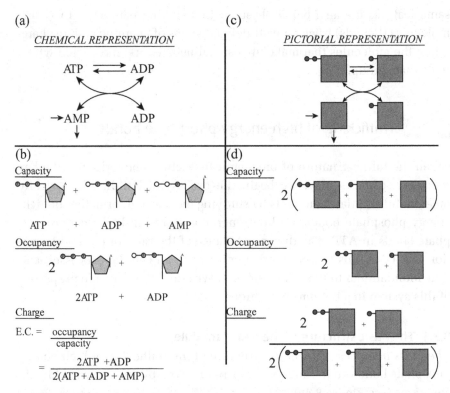

Figure 8.4 Representation of the exchange of high-energy phosphate bonds among the adenosine phosphates. (a) The chemical reactions. (b) The molecules with open circles showing the "vacant" places for high-energy bonds. The capacity to carry high-energy phosphate bonds, the occupancy of high-energy bonds, and the energy charge are shown. (c) The reaction schema of (a) in pictorial form. The solid squares represent AMP and the solid circles the high-energy phosphate bonds. (d) The same concepts as in (b) represented in pictorial form.

The use and formation of ATP is about 10 mM/min, resulting in $k_{\text{use}} = 6.25/\text{min}$ and $k_{\text{form}} = 25/\text{min}$.

The net rate for the redistribution of high-energy bonds is

$$v_{\text{distr}} = v_{+\text{distr}} - v_{-\text{distr}} \tag{8.6}$$

$$= k_{\text{distr}}\text{ADP}^2 - k_{-\text{distr}}\text{ATP} \times \text{AMP}$$

$$= k_{\text{distr}}(\text{ADP}^2 - \text{ATP} \times \text{AMP}/K_{\text{distr}}),$$

where K_{distr} is about unity and k_{distr} is large (we set it here at 1000/min). These numerical values for the concentrations and the kinetic constants lead to a steady state. Note that there is another steady-state solution of ATP = ADP = 0 and AMP = 2.1, which is a "ground" state (i.e., no charge) for the system. The next chapter discusses multiple steady states in more detail.

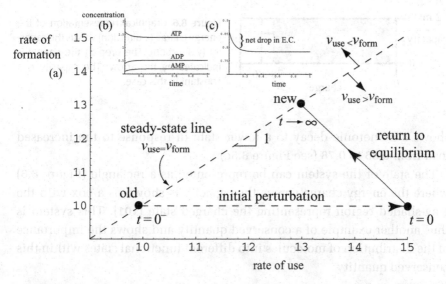

Figure 8.5 Dynamic responses for Eqs (8.2)–(8.5). (a) The phase portrait for the rates of use and formation of ATP. (b) The concentrations of ATP, ADP, and AMP. (c) The Atkinson's energy charge (Eq. (8.7)).

Dynamic simulations The dynamic mass balances can be simulated for a change in the initial conditions or in the numerical values of the parameters. Here, we simulate the response of this system to a 50% increase in the rate of ATP use. At time zero, we have the network in a steady state and we change k_{use} from 6.25/min to $1.5 \times 6.25 = 9.375$/min, and the rate of ATP use instantly becomes 15 mM/min.

The dynamic response shows that all the concentrations change with ATP concentration dropping with the increased load (Figure 8.5). The response of the system is perhaps best visualized in Figure 8.5a, showing the phase portrait of the rate of ATP use versus ATP formation. Prior to the increased load, the system is on the 45° line and at time zero it is instantly imbalanced and moved into the region where more ATP is used than formed (15 versus 10 mM/min). From this initial perturbation the response of the system is to move directly towards the 45° line to regain balance between ATP use and formation.

Pooling and interpretation Since AMP is not being synthesized and degraded, the sum of ATP + ADP + AMP, or the capacity to carry high-energy phosphate bonds, is a constant; see Figure 8.4c. The Atkinson's energy charge

$$\text{E.C.} = \frac{2\text{ATP} + \text{ADP}}{2(\text{ATP} + \text{ADP} + \text{AMP})} = \frac{\text{occupancy}}{\text{capacity}} \qquad (8.7)$$

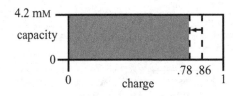

Figure 8.6 Graphical representation of the energy charge (x-direction) versus the capacity (y-direction). The drop in the charge is indicated by the arrow. The capacity is a constant in this case.

shows a monotonic decay to a lower state in response to the increased load from 0.86 to 0.78 (see Figure 8.5c).

The state of the system can be represented as a rectangle (Figure 8.6) where the energy charge versus the capacity is shown as a box with the gray-shaded region representing the charged state [101]. This system is thus another example of a conserved quantity and shows the importance of the distribution of molecules into different functional states within this conserved quantity.

8.3.2 Buffering the energy charge

Reaction mechanism   In many situations, there is a buffering effect on the energy charge by a coupled carrier of high-energy bonds. This exchange is

$$ATP + B \rightleftharpoons ADP + BP, \tag{8.8}$$

where the buffering molecule B picks up the high-energy phosphate group through a fast equilibrating reaction.

Example of buffer molecules   In Eq. (8.8), B represents a phosphagen, which is a compound containing a high-energy phosphate bond that is used as energy storage to buffer the ATP/ADP ratio. The most well-known phosphagen is creatine, which is found in the muscles of mammals. Marine organisms have other phosphagens (arginine, taurocyamine, glycocyamine), while earthworms use lombricine [119].

Buffering   When the reaction in Eq. (8.8) is at equilibrium we have

$$k_{buff}ATP \times B = k_{-buff}ADP \times BP. \tag{8.9}$$

This equation can be rearranged as

$$4K_{buff} = BP/B, \tag{8.10}$$

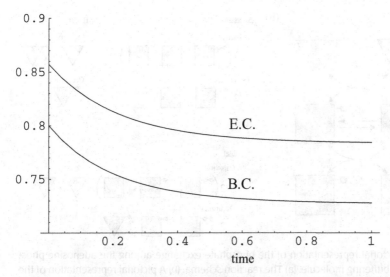

Figure 8.7 Dynamic responses for Eq. (8.5) with the buffering effect (Eq. (8.8)). The Atkinson's energy charge (Eq. (8.7)) and the buffer charge are shown as a function of time. $B_{tot} = 10$, $K_{buff} = 1$, and $k_{buff} = 1000$. All other conditions are as in Figure 8.5, i.e., we simulate the response to a 50% increase in k_{use}.

where $ATP/ADP = 1.6/0.4 = 4$ in the steady state and $K_{buff} = k_{buff}/k_{-buff}$. If the buffering molecule is present in a constant amount, then

$$B_{tot} = B + BP. \tag{8.11}$$

We can rearrange these equations to be

$$\frac{BP}{B_{tot}} = \frac{4K_{buff}}{4K_{buff} + 1}. \tag{8.12}$$

In this equation, B_{tot} is the capacity of the buffer to carry the high-energy phosphate bond, whereas BP/B_{tot} is the energy charge of the buffer. We note that the value of K_{buff} is a key variable. If $K_{buff} = 1/4$ then the buffer is half charged at equilibrium, whereas if $K_{buff} = 1$ then the buffer is 80% charged. Thus, this numerical value (a thermodynamic quantity) is key and will specify the relative charge on the buffer and the adenosine phosphates. The effect of K_{buff} can be determined through simulation.

Simulation   The response of the adenosine phosphate system can be simulated in the presence of a buffer. We choose the parameters as $B_{tot} = 10$ mM, $K_{buff} = 1$, $k_{buff} = 1000/min$, and all other conditions as in Figure 8.5. The results of the simulation are shown in Figure 8.7. The time response of the energy charge is shown, along with the buffer charge BP/B_{tot}. We see that the fast response in the energy charge is now slower, as the initial reaction is buffered by release of the high-energy bonds that

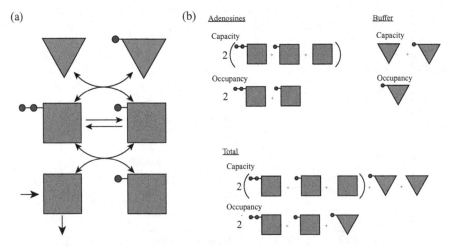

Figure 8.8 Pictorial representation of the phosphate exchange among the adenosine phosphates and a buffering molecule. (a) The reaction schema. (b) A pictorial representation of the molecules, their charged states, and the definition of pooled variables.

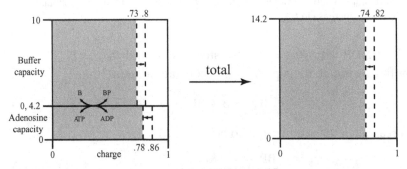

Figure 8.9 The representation of the energy and buffer charge versus the capacity (in millimoles on y-axis). The lumping of the two quantities into "overall" quantities is illustrated. The case considered corresponds to the simulation in Figure 8.7.

are bound to the buffer. The overall change in the energy charge is the same: it goes from 0.86 to 0.78. The charge of the buffer drops from 0.80 to 0.73 at the same time.

Pooling and interpretation A pictorial representation of the phosphate buffering is given in Figure 8.8. Here, a generalized definition of the overall phosphate charge is

$$\text{overall charge} = \frac{\text{overall occupancy}}{\text{overall capacity}} = \frac{2\text{ATP} + \text{ADP} + \text{BP}}{2(\text{ATP} + \text{ADP} + \text{AMP}) + \text{BP} + \text{B}}.$$

This combined charge system can be represented similarly to the representation in Figure 8.6. Figure 8.9 shows a stacking of the buffer and adenosine phosphate capacity versus their charge. The total capacity to carry high-energy bonds is now 14.2 mM. The overall charge is 0.82

(or 11.64 mM concentration of high-energy bonds) in the system before the perturbation. The increased load brings the overall charge down to 0.74.

8.3.3 Open system: long-term adjustment of the capacity

Inputs and outputs Although the rates of formation and degradation of AMP are low, their effects can be significant. These fluxes will determine the total amount of the adenosine phosphates and thus their capacity to carry high-energy bonds. The additional elementary rate laws needed to account for the rate of AMP formation and drain are

$$v_{\text{form,AMP}} = b_1, \quad v_{\text{drain}} = k_{\text{drain}}\text{AMP}, \tag{8.13}$$

where b_1 is the net synthesis rate of AMP. The numerical values used are $b_1 = 0.03$ mM/min and $k_{\text{drain}} = (0.03 \text{ mM/min})/(0.1 \text{ mM}) = 0.3$ mM/min.

Simulation This system has a biphasic response for these values of the kinetic constants. We can start the system in a steady state at $t = 0^-$ and simulate the response for increasing the ATP load by shifting the value of k_{use} by 50% at $t = 0$, as before. The initial rapid response is similar to what is shown in Figure 8.5a, where the concentration of ATP drops in response to the load and the concentrations of ADP and AMP rise. This initial response is followed by a much slower response where all three concentrations drop.

Interpretation This biphasic response can be examined further by look-ing at dynamic phase portraits of key fluxes (Figure 8.10) and key pools (Figure 8.11).

- First, we examine how the system balances the use of ATP (v_{use}) with its rate of formation (v_{form}); see Figure 8.10a. At $t = 0$ the system is at rest at $v_{\text{use}} = v_{\text{form}} = 10.0$ mM/min. Then the system is perturbed by moving the ATP drain, v_{use}, to 15.0 mM/min, as before. The initial response is to increase the formation rate of ATP to about 13 mM/min with the simultaneous drop in the use rate to about the same number, due to a net drop in the concentration of ATP during this period. The rate of ATP use and formation is approximately the same at this point in time. Then, during the slower response time, the use and formation rates of ATP are similar and the system moves along the 45° line to a new steady-state point at 6.67 mM/min.
- The slow dynamics are associated with the inventory of the adenosine phosphates (ATP + ADP + AMP). The AMP drain can be graphed ver-sus the ATP use; see Figure 8.10b. Initially, the AMP drain increases rapidly as the increased ATP use leads to ADP buildup that gets converted into AMP by adenylate kinase (v_{distr}). The AMP drain then

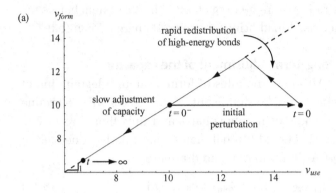

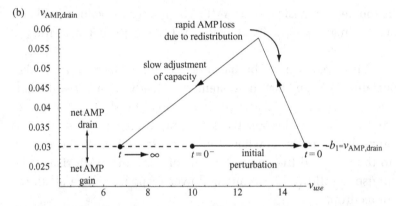

Figure 8.10 Dynamic phase portraits of fluxes for the simulation of the adenosine phosphate system with formation and drain of AMP (Eq. (8.13)). (a) The ATP use v_{use} versus the ATP formation rate v_{form}. (b) The ATP use v_{use} versus the AMP drain v_{drain}.

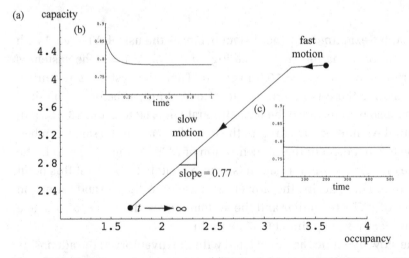

Figure 8.11 The energy charge response. (a) Dynamic phase portrait of $2ATP + ADP$ versus $2(ATP + ADP + AMP)$. (b) The response of E.C. on the fast time scale. (c) The response of E.C. on the slow time scale.

drops and sets at the same rate to balance the formation rate, set at 0.03 mM/min.

- We can graph the occupancy against the capacity (Figure 8.11a). During the initial response, the occupancy moves while the capacity is a constant. Then, during the slower phase, the two move at a constant ratio. This gives a biphasic response of the energy charge (Figure 8.11b and c). It rapidly changes from 0.86 to about 0.77 and then *stays a constant*. The energy charge is roughly a constant even though all the other concentrations are changing.

This feature of keeping the energy charge a constant while the capacity is changing has a role in a variety of physiological responses, from blood storage to the ischemic response in the heart. Note that this property is a stoichiometric one; no regulation is required to produce this effect.

8.4 Charging and recovering high-energy bonds

Reaction mechanism   As discussed in Section 8.2, most catabolic pathways generate energy (and other metabolic resources) in the form of activated (or charged) carrier molecules. Before energy can be extracted from a compound, it is typically spent (a biological equivalent of "it takes money to make money"). This basic structure shown in Figure 2.5 is redrawn in Figure 8.12a, where one ATP molecule is used to "charge" a substrate (x_1) with one high-energy bond to form an intermediate (x_2). This intermediate is then degraded through a process wherein two ATP molecules are synthesized and an inorganic phosphate is incorporated. The net gain of ATP is one for every x_1 metabolized, and this ATP molecule can then be used to drive a process (v_{load}) that uses an ATP molecule. The trafficking of high-energy phosphate bonds is shown pictorially in Figure 8.12b.

The dynamic mass balances   The dynamic mass balance equations that describe this process are

$$\frac{dx_1}{dt} = b_2 - v_1, \tag{8.14}$$

$$\frac{dx_2}{dt} = v_1 - v_2, \tag{8.15}$$

$$\frac{d\,\text{ATP}}{dt} = -(v_1 + v_{load}) + 2v_2 + v_{distr}, \tag{8.16}$$

$$\frac{d\,\text{ADP}}{dt} = (v_1 + v_{load}) - 2v_2 - 2v_{distr}, \tag{8.17}$$

$$\frac{d\,\text{AMP}}{dt} = b_1 - v_{drain} + v_{distr}. \tag{8.18}$$

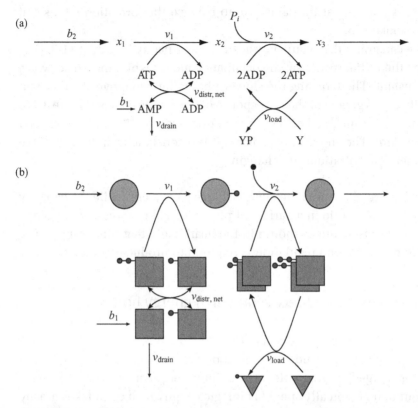

Figure 8.12 Coupling of the adenosine phosphates with a skeleton metabolic pathway. (a) The reaction map. (b) A pictorial view of the molecules emphasizing the exchange of the high-energy phosphate group (solid circle). The square is AMP. The rate laws used are $b_1 = 0.03\,\text{mM/min}$; $b_2 = 5\,\text{mM/min}$; $k_{\text{drain}} = b_1/0.1$; $k_{\text{load}} = 5/1.6$; $k_1 = 5/1.6$; $k_2 = 5/0.4$.

To integrate the reaction schema in Figure 8.4a with this skeleton pathway, we have replaced the use rate of ATP (v_{use}) with $v_1 + v_{\text{load}}$ and the formation rate of ATP (v_{form}) with $2v_2$.

Simulation The flow of substrate into the cell, given by b_2, will be set to 5 mM/min in the simulation to follow to set the gross ATP production at 10 mM/min. The response of this system can be simulated to a 50% increase in the ATP load parameter, as in previous examples. The difference from the previous examples here is that the net ATP production rate is 5 mM/min. The time responses of the concentrations and fluxes are shown in Figures 8.13 and 8.14.

Interpretation We can make the following observations from this dynamic response:

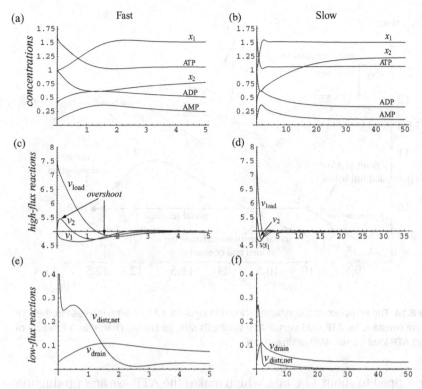

Figure 8.13 The response of the systems shown in Figure 8.12 to a 50% increase in the ATP load rate constant. (a, b) Dynamic response of the concentrations on a fast and slow time scale. (c, d) Dynamic response of the main fluxes on a fast and slow time scale. (e, f) Dynamic response of the AMP determining fluxes on a fast and slow time scale. Parameter values are the same as in Figure 8.12.

- The concentrations move on two principal time scales (Figure 8.13a and b): a fast time scale that is about 3 to 5 min, and a slower time scale that is about 50 min. ATP and x_1 move primarily on the fast time scale, whereas ADP, AMP, and x_2 move on the slower time scale.

- Initially, v_{load} increases sharply and v_2 increases and v_1 decreases to meet the increased load. The three high flux reactions v_1, v_2, and v_{load} restabilize at about 5 mM/min after about a 3 to 5 min time frame, after which they are closely, but not fully, balanced (Figure 8.13c and d).

- The dynamic phase portrait, Figure 8.14a, shows that the overall ATP use $(v_1 + v_{load})$ quickly moves to about 12.5 mM/min while the production rate $(2v_2)$ is about 10 mM/min. Following this initial response, the ATP use drops and the ATP synthesis rate increases to move towards the 45° line. The 45° line is not reached. After 0.1 min, v_2 starts to drop and the system moves somewhat parallel to the 45° line until 1.5 min have passed. At this time the ATP concentration has

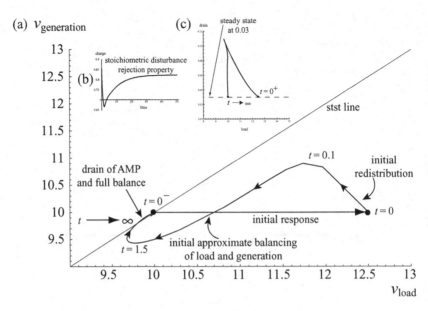

Figure 8.14 The response of the systems shown in Figure 8.12 to a 50% increase in the ATP load rate constant. (a) ATP load versus ATP synthesis rate. (b) Energy charge as a function of time. (c) ATP load versus AMP drainage rate.

dropped to about 1.06 mM, which makes the ATP use and production rate approximately balanced. Following this point, both the use and production rate increase slowly and return the system back to the initial point where both have a value of 10 mM/min. Since the input rate of x_1 is a constant, the system has to return to the initial state.

- AMP initially increases leading to a net drain of AMP from the system. This drain unfolds on a long time scale, leading to a net flux through the adenylate kinase that decays on the slower time scale. The effects of AMP drainage can be seen in the flux phase portrait in Figure 8.14c. Initially, the AMP drain increases as the ATP usage drops close to its eventual steady state. Then the vertical motion in the phase portrait shows that there is a slower motion in which the ATP usage does not change much but the AMP drainage rate drops to match its input rate at 0.03 mM/h.

- The dynamic response of the energy charge (Figure 8.14b) shows that it drops on the faster time scale from an initial value of 0.86 to reach a minimum of about 0.67 at about 1.5 min. This initial response results from the increase in the ATP load parameter of 50%. After this initial response, the energy charge increases on the slower time scale to an eventual value of about 0.82.

- Notice that this secondary response is not a result of a regulatory mechanism, but is a property that is built into the stoichiometric structure and the values of the rate constants that lead to the time-scale separation.

8.5   Summary

➤ Most biochemical reactions are bilinear. Six of the seven categories of enzymes catalyze bilinear reactions.

➤ The bilinear properties of biochemical reactions lead to complex patterns of exchange of key chemical moieties and properties. Many such simultaneous exchange processes lead to a "tangle of cycles" in biochemical reaction networks.

➤ Skeleton (or scaffold) dynamic models of biochemical processes can be carried out using dynamic mass balances based on elementary reaction representations and mass action kinetics.

➤ Complex kinetic models are built in a bottom-up fashion, adding more details in a stepwise fashion, making sure that every new feature is consistently integrated. This chapter demonstrated a three-step analysis of the ATP cofactor sub-network and then its integration to a skeleton ATP-generating pathway.

➤ Once dynamic network models are formulated, the perturbations to which we simulate their responses are in fluxes, typically the exchange and demand fluxes.

➤ A recurring theme is the formation of pools and the state of those pools in terms of how their total concentration is distributed among its constituent members.

➤ Some dynamic properties are a result of the stoichiometric structure and do not result from intricate regulatory mechanisms or complex kinetic expressions.

## CHAPTER 9

Regulation as elementary phenomena

In Chapter 8 we demonstrated that the dynamic states of biochemical reaction networks can be characterized by elementary reactions. We now show how the phenomena of regulation can be described and simulated in the same framework. After summarizing genetic and biochemical mechanisms that regulate enzymes, we discuss the basic phenomenological concepts associated with regulation. To demonstrate some of dynamic consequences of regulation, we then discuss in mathematical detail the effects of local feedback inhibition and feedforward activation. Finally, we set up simulation models of regulation of a prototypical biosynthetic pathway. The chemical reactions that underlie the regulatory steps are identified and their kinetic properties are estimated. These reactions are then added to the scaffold formed by the basic mass action kinetic description of the network of interest to simulate the effects of the regulation. This approach will then be applied to realistic situations in Part IV.

9.1 Regulation of enzymes

Many factors regulate enzymes, their concentrations, and their catalytic activities. We describe four different mechanisms here (Figure 9.1).

- *Regulation of gene expression:* the transcription of genes is regulated in an intricate way. Many proteins, called transcription factors, bind to the promoter region of a gene. Their binding can *induce* or *repress* gene expression. Metabolites often determine the active states of these regulatory proteins.
- *Interconversion:* regulated enzymes can exist in many functional states. As we saw in Chapter 5, a regulatory enzyme can be modeled as existing in two conformations: catalytically active and catalytically

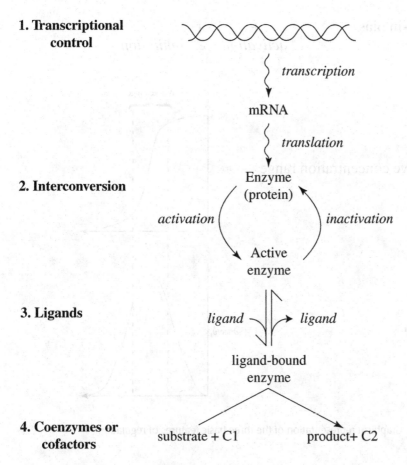

1. Transcriptional control

transcription

mRNA

translation

2. Interconversion

Enzyme (protein)

activation   *inactivation*

Active enzyme

3. Ligands

ligand   *ligand*

ligand-bound enzyme

4. Coenzymes or cofactors   substrate + C1   product+ C2

Figure 9.1 Four levels of regulation of enzymes: gene expression, interconversion, ligand binding, and cofactor availability. C1 and C2 represent a cofactor or a coenzyme and its two different states (charged and discharged).

inactive. Often, regulated enzymes are chemically modified through phosphorylation, methylation, or acylation to convert them between inactive and active states.

- *Binding by ligands:* small molecules can bind to regulatory enzymes in an allosteric binding site (see Chapters 5 and 14). Such binding can promote the relaxed (R) or taut (T) state of the enzyme, leading to a "tug of war" among its states.
- *Cofactor and coenzyme availability:* as detailed at the beginning of Chapter 8, enzymes rely on "accessory molecules" for their function. Thus, the availability of such molecules determines the functional state of the enzyme.

These are four genetic and biochemical mechanisms by which regulation of enzyme catalytic activity is exerted. We will now describe the dynamic consequences of such regulatory actions.

1. Built-in bias   +   or   −

                          activation   *inhibition*

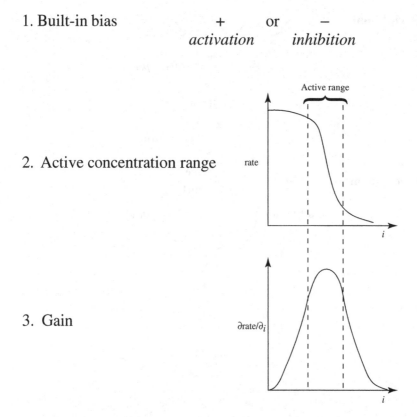

2. Active concentration range

3. Gain

Figure 9.2 Graphical representation of the three basic features of regulation.

9.2   Regulatory signals: phenomenology

Basic features of a regulatory event   The regulatory action of compounds is characterized by three important measures: bias, active range, and sensitivity. These features are illustrated in Figure 9.2 and are detailed as follows:

- First, there is a built-in bias; the regulation is either negative (inhibition) or positive (activation).
- Second, there is a range of concentrations over which the signal is active. The measure of this concentration range is the dissociation constant for the regulatory molecule.
- Third is the sensitivity of the flux to changes in the concentration of the regulator, or the "gain" of the regulation.

Network topology   Regulation can be exerted "close" to the formation of the regulatory molecule, such as feedback activation of an enzyme by its product, or the regulator can be controlling "far" away from its site formation, as is the case with citrate inhibition of PFK or for amino acids feedback inhibiting the first reaction in the sequence that leads to their formation; see Figure 9.7.

Physiological roles   From a network perspective, there are two overall physiological functions of regulatory signals:

- To overcome any disturbances in the cellular environment. This function is the "disturbance rejection" problem. A cell or an organism experiences a change in an environmental parameter but wants to maintain its homeostatic state.
- To drive a network from one state to the next; that is, change the homeostatic state. This need is broad and often encountered from the need to change from a nongrowing state to a growing one in response to the availability of a nutrient or to the need for a precursor cell to initiate differentiation to a new state. In control theory, this is known as the "servo" problem.

9.3   The effects of regulation on dynamic states

We now quantitatively assess the effects of regulatory signals on network dynamic states, representing the most mathematically difficult material in this book.

Basic mathematical features   To examine the qualitative effects of signals on network dynamics, let us examine the simple scheme

$$\xrightarrow{v_1(x)} X \xrightarrow{v_2(x)} , \tag{9.1}$$

where the concentration of metabolite X, x, directly influences the rates of its own formation, $v_1(x)$, and degradation, $v_2(x)$. The following discussion is illustrated in Figure 9.3.
    The dynamic mass balance on X is

$$\frac{dx}{dt} = v_1(x) - v_2(x) \tag{9.2}$$

that in a linearized form is

$$\frac{dx'}{dt} = \left(\frac{\partial v_1}{\partial x} - \frac{\partial v_2}{\partial x} \right) x' = \lambda x', \tag{9.3}$$

where λ is the "net" rate constant. The value of λ determines the rate of response of this system to changes in the concentration of X.

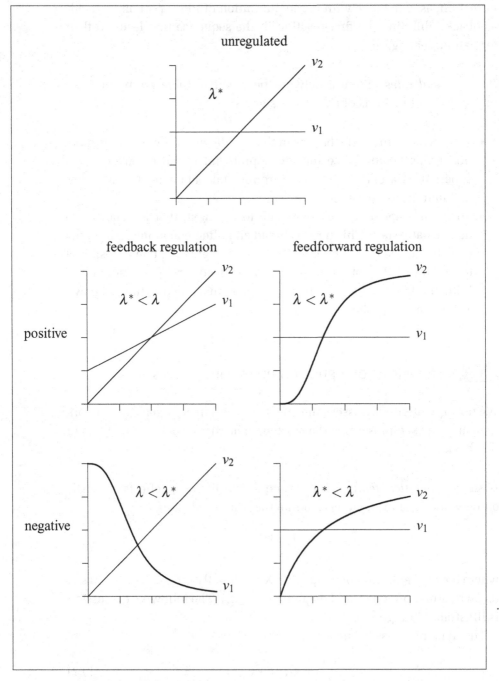

Figure 9.3 Qualitative effects of regulatory signals. The graphs show the dependency of flux on the concentration of X.

Measures of systemic effects The quantity λ is a combination of the time constants for the individual reactions. It is thus a "systems" property rather than property of a single component or a link in the network. It is equivalent to the eigenvalue of this one-dimensional system. Eigenvalues are systems quantities that are combinations of the properties of the individual components and links. These combinations become complicated as the size of a networks grows.

Local regulation We examine the various scenarios that can arise when regulation is added on top of the natural mass action trend:

- *The unregulated situation:* When no regulation is exerted, the eigenvalue that describes the dynamics is given by

$$\lambda = \frac{\partial v_1}{\partial x} - \frac{\partial v_2}{\partial x} = 0 - k_2 = \lambda^* < 0 \qquad (9.4)$$

  if we assume that the turnover rate is first order and the formation rate is zeroth order. The unregulated situation corresponds to elementary mass action kinetics overlaid on the stoichiometric structure.
- *Feedback inhibition:* If the formation rate is feedback inhibited, the eigenvalue becomes more negative than for the unregulated situation since now $\partial v_1/\partial x < 0$, and hence the time constant is faster. The feedback inhibition, therefore, augments the mass action trend and supports the natural dynamics of the system.
- *Feedback activation:* Feedback activation does just the opposite. Now $\partial v_1/\partial x > 0$ and this signal thus tends to counter the natural dynamics of the system. If the signal is sufficiently strong, then the eigenvalue can become zero or positive, creating an instability that is reflected as multiple steady states.
- *Feedforward inhibition:* This signal tends to reduce $\partial v_2/\partial x$ and hence tends to move λ closer to zero, making instability more likely. Feedforward inhibition counters the mass action trend and is classified as a destabilizing signal. Note that saturation kinetics are of this type and they are observed to create instabilities in several models [91, 106, 118].
- *Feedforward activation:* Feedforward activation increases the magnitude of λ and thus supports the mass action trend.

Regulatory principles Regulatory signals can either support or antagonize the mass action trend in a network. Thus, we arrive at the following principles:

1. Local negative feedback and local positive feedforward controls support the mass action trend and are stabilizing in the sense that they try to maintain the intrinsic dynamic properties of the stoichiometric structure.
2. Local positive feedback and local negative feedforward controls counteract the mass action trend and can create instabilities. Many of the creative functions associated with metabolism can be attributed to these control modes. These signals allow the cell to behave in apparent defiance to the laws of mass action and stoichiometric trends.

Regulators that act "far" from their site of formation can induce dynamic instabilities even when the signal supports the mass action trend. There is a limit on the extent of stabilization or support of homeostasis achievable.

Measuring the dynamic effects of regulation Equation (9.4) may be rewritten as

$$\lambda = -\frac{\partial v_2}{\partial x}\left(1 - \frac{\partial v_1/\partial x}{\partial v_2/\partial x}\right) \tag{9.5}$$

$$\approx -\frac{\partial v_2}{\partial x}\left(1 \pm \frac{t_{\text{turnover}}}{t_{\text{regulation}}}\right) \tag{9.6}$$

if the time constants indicated are good estimates of the corresponding partial derivatives. The dimensionless ratio

$$a = \frac{t_{\text{turnover}}}{t_{\text{regulation}}} \tag{9.7}$$

thus characterizes local regulatory signals. If a is less than unity or on the order of unity, the regulation is dynamically about as important as the natural turnover time. However, if it significantly exceeds unity, one would expect that dynamics then would become dominated by the regulatory action.

9.4   Local regulation with Hill kinetics

We will now look at specific examples to quantitatively explore the concepts introduced in the last section. We will use Hill-type rate laws that are the simplest mathematical forms for regulated reactions. Even for the simplest case, the algebra becomes a bit cumbersome.

9.4.1 Inhibition

We can look quantitatively at the effects of local feedback inhibition. We can consider specific functional forms for v_1 and v_2 in Eq. (9.2):

$$v_1(x) = \frac{v_m}{1 + (x/K)^2} \quad \text{and} \quad v_2(x) = kx, \tag{9.8}$$

where the production rate is a Hill-type equation with $v = 2$ and the removal is an elementary first-order equation. The dynamic mass balance is

$$\frac{dx}{dt} = \frac{v_m}{1 + (x/K)^2} - kx. \tag{9.9}$$

The dynamics of this simple system can be analyzed to determine the dynamic effects of the feedback inhibition.

The steady state The steady-state equation, $v_1 = v_2$, for this network is a cubic equation:

$$\left(\frac{x}{K}\right)^3 + \frac{x}{K} - \frac{v_m}{kK} = 0. \tag{9.10}$$

Introducing a dimensionless concentration $\chi = x/K$ we have that

$$\chi^3 + \chi - a = 0, \tag{9.11}$$

where $a = v_m/kK$. This equation has one real root

$$\chi_{ss} = \frac{\sqrt[3]{9a + \sqrt{3}\sqrt{27a^2 + 4}}}{3\sqrt[3]{\frac{2}{3}}} - \frac{\sqrt[3]{\frac{2}{3}}}{\sqrt[3]{9a + \sqrt{3}\sqrt{27a^2 + 4}}}. \tag{9.12}$$

The steady-state level of x, relative to K, is dependent on a single parameter, a, as shown in Figure 9.4. We see that

$$a = \frac{1/k}{K/v_m} = \frac{t_{turnover}}{t_{regulation}}. \tag{9.13}$$

Determining the eigenvalue By defining $\tau = kt$ we have

$$\frac{d\chi}{d\tau} = \frac{a}{1 + \chi^2} - \chi \quad \text{and thus} \quad \chi_{ss} = \frac{a}{1 + \chi_{ss}^2}. \tag{9.14}$$

The eigenvalue can be computed thus:

$$\lambda = \frac{\partial}{\partial \chi}\left[\frac{a}{1 + \chi^2} - \chi\right]_{ss} = -\frac{2a\chi_{ss}}{(1 + \chi_{ss}^2)^2} - 1 = -\frac{2\chi_{ss}^2}{1 + \chi_{ss}^2} - 1. \tag{9.15}$$

The eigenvalue thus changes from a negative unity (when $x_{ss} \ll K$ and the regulation is not felt) to a negative three (when $x_{ss} \gg K$ where regulation is strong).

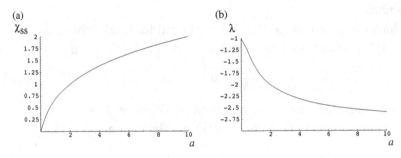

Figure 9.4 Regulation by local feedback inhibition. (a) The dependency of the steady state concentration, $\chi_{ss} = x/K$, on $a = v_m/kK$ by solving Eq. (9.11). (b) The eigenvalue as a function of a computed from Eq. (9.15).

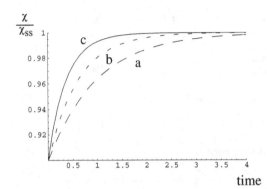

Figure 9.5 Regulation by local feedback inhibition. The dynamic responses of $d\chi/d\tau = a/(1 + \chi^2) - \chi$ for: (a) long dashes, $a = 0.1$ ($\chi_{ss} = 0.1$, $\lambda = -1.02$); (b) short dashes, $a = 1.0$ ($\chi_{ss} = 0.68$, $\lambda = -1.64$); (c) solid line, $a = 10$ ($\chi_{ss} = 2$, $\lambda = -2.6$). The initial conditions for each case are $0.9\chi_{ss}$. Each curve is graphed as $\chi(t)/\chi_{ss}$.

Dynamic simulations The dynamic effects of the feedback regulation can be simulated; Figure 9.5. Three cases are considered by changing the single dimensionless parameter a. As a increases, the regulation is tighter and the value of λ increases, thus making the approach to steady state faster.

9.4.2 Activation

We can look quantitatively at the effects of local feedback activation. We can consider specific functional forms for v_1 and v_2 in Eq. (9.2):

$$v_1(x) = v_m \frac{1 + \alpha(x/K)^\nu}{1 + (x/K)^\nu} \quad \text{and} \quad v_2(x) = kx \tag{9.16}$$

with a dynamic mass balance

$$\frac{dx}{dt} = v_m \frac{1 + \alpha(x/K)^\nu}{1 + (x/K)^\nu} - kx. \tag{9.17}$$

We can make this equation dimensionless using the same dimensionless variables as above:

$$\frac{d\chi}{d\tau} = a\frac{1+\alpha\chi^{\nu}}{1+\chi^{\nu}} - \chi. \tag{9.18}$$

The steady state   In a steady state, the dynamic mass balance becomes

$$a\frac{1+\alpha\chi_{ss}^{\nu}}{1+\chi_{ss}^{\nu}} = \chi_{ss}. \tag{9.19}$$

Determining the eigenvalue   The eigenvalue can be determined by the linearization of Eq. (9.18):

$$\lambda = \frac{\nu a(\alpha-1)\chi_{ss}^{\nu-1}}{(1+\chi_{ss}^{\nu})^2} - 1; \tag{9.20}$$

since $\alpha > 1$, the first term in Eq (9.20) is positive. This leads to the possibility that the eigenvalue is zero. This condition in turn leads to a situation where one can have multiple steady states, as will now be demonstrated.

Existence of multiple steady states   We are looking for conditions where Eqs (9.19) and (9.20) are simultaneously zero; that is, the steady-state condition and a zero eigenvalue. The two equations can be combined by multiplying Eq. (9.20) by χ and adding the equations together. After rearrangement, the equations become

$$z^2 + [(1-\nu) + (1+\nu)/\alpha]z + 1/\alpha = 0, \quad z = \chi^{\nu}. \tag{9.21}$$

This equation can only have a real positive solution if

$$\alpha > \alpha_{\min} = \left(\frac{1+\nu}{1-\nu}\right)^2. \tag{9.22}$$

If α exceeds this minimum value, the steady-state equation will have multiple solutions for χ for a range of values for a.

Region of multiple steady states in the parameter plane   If α exceeds its minimum value, multiple steady states are possible. Equation (9.21) will have two roots, $\chi_{1,ss}$ and $\chi_{2,ss}$. These roots depend on two parameters, a and α. To compute the relationships between a and α when the eigenvalue is zero, we can first specify α and compute $\chi_{1,ss}$ and $\chi_{2,ss}$ and then compute the two corresponding values for a_1 and a_2:

$$a_1 = \chi_{1,ss}\left(\frac{1+\chi_{1,ss}^{\nu}}{1+\alpha\chi_{1,ss}^{\nu}}\right), \tag{9.23}$$

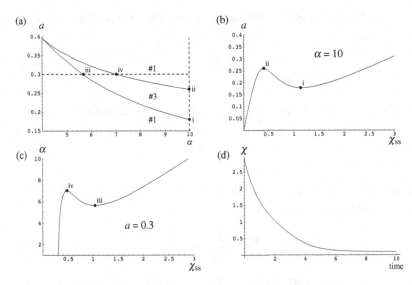

Figure 9.6 Regulation by local feedback activation. (a) The plane of a and α and the regions in the plane that can have one or three steady states. (b) The steady-state concentration χ_{ss} as a function of a for $\alpha = 10$. (c) The steady-state concentration χ_{ss} as a function of α for $a = 0.3$. (d) The dynamic response from an initial parameter set of $(a, \alpha) = (0.3, 10)$ to $(a, \alpha) = (0.1, 10)$. $\nu = 3$. The points indicated by i, ii, iii, iv are selected critical points where the steady-state solution "turns around," thus demarcating the region in the parameter space where multiple steady-state solutions exist.

and the same for a_2 as a function of $\chi_{2,ss}$. The results are shown in Figure 9.6a.

Computing the multiple steady-state values for the steady-state concentration For a fixed value of α we can vary a along a line in the parameter plane and compute the steady-state concentration. Since one cannot solve the steady-state solution explicitly for χ_{ss}, this is hard to do; however, one can rearrange it as

$$a = \chi_{ss} \left(\frac{1 + \chi_{ss}^{\nu}}{1 + \alpha \chi_{ss}^{\nu}} \right) \tag{9.24}$$

and plot a versus χ to get the same results. The limit, or turnaround point, corresponds to substituting the roots of Eq. (9.21) into Eq. (9.24). The resulting graphs are shown in Figure 9.6b. Similarly, one can fix a and compute the relationship between χ_{ss} and α; see Figure 9.6c.

Simulating the dynamic response to a critical change in the parameter values We can simulate the dynamic response of this loop to changes in the parameter values. As an example, we chose $\alpha = 10$ and change a from an initial value of 0.3 to 0.1 at time zero. These two points lie on each

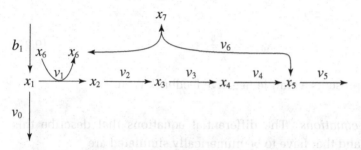

Figure 9.7 A schematic of a prototypical feedback loop for a biosynthetic pathway. The end product of the pathway, x_5, feedback inhibits the flux into the pathway.

side of the region of three steady states (Figure 9.6a). The results from the dynamic simulation are shown in Figure 9.6d.

9.5   Feedback inhibition of pathways

The local regulation of a reaction by its reactants or products has just been analyzed. We now turn our attention to situations where a reaction is regulated by a compound that is far away in the network and to how to build simulation models of such regulatory phenomena. We can use simulation and case studies to determine the effectiveness of the regulatory action. We will now use elementary reactions to describe the regulatory mechanism.

Feedback regulation in a biosynthetic pathway In a biosynthetic pathway, the first reaction is often inhibited by the end product of the pathway (Figure 9.7). A metabolic intermediate x_1 (called a biosynthetic precursor – there are 12 key such molecules in metabolism) is being formed and degraded as

$$\xrightarrow{b_1} x_1 \xrightarrow{k_0} . \tag{9.25}$$

Then, if an enzyme x_6 is expressed, x_1 can be converted to x_2:

$$x_1 + x_6 \xrightarrow{k_1} x_2 + x_6, \tag{9.26}$$

which is followed by a series of reactions

$$x_2 \xrightarrow{k_2} x_3 \xrightarrow{k_3} x_4 \xrightarrow{k_4} x_5 \xrightarrow{k_5} \tag{9.27}$$

to form x_5, which is an end product of the pathway. The end product does feedback inhibit the enzyme x_6 by binding to it and converting it into an inactive form:

$$x_6 + x_5 \underset{k_{-6}}{\overset{k_6}{\rightleftharpoons}} x_7. \tag{9.28}$$

This system represents a simple negative feedback loop.

The dynamic equations   The differential equations that describe this feedback loop and that have to be numerically simulated are

$$\frac{dx_1}{dt} = b_1 - v_0 - v_1 = b_1 - k_0 x_1 - k_1 x_6 x_1, \tag{9.29}$$

$$\frac{dx_2}{dt} = \quad v_1 - v_2 \quad = k_1 x_6 x_1 - k_2 x_2, \tag{9.30}$$

$$\frac{dx_3}{dt} = \quad v_2 - v_3 \quad = k_2 x_2 - k_3 x_3, \tag{9.31}$$

$$\frac{dx_4}{dt} = \quad v_3 - v_4 \quad = k_3 x_3 - k_4 x_4, \tag{9.32}$$

$$\frac{dx_5}{dt} = v_4 - v_5 - v_6 = k_4 x_4 - k_5 x_5 - (k_6 x_5 x_6 - k_{-6} x_7), \tag{9.33}$$

$$\frac{dx_6}{dt} = \quad -v_6 \quad = -k_6 x_5 x_6 + k_{-6} x_7, \tag{9.34}$$

$$\frac{dx_7}{dt} = \quad v_6 \quad = k_6 x_5 x_6 - k_{-6} x_7. \tag{9.35}$$

The sum of the last two differential equations gives us the mass balance in the enzyme:

$$x_6 + x_7 = e_t, \tag{9.36}$$

where e_t is the total amount of enzyme.

The steady-state equations   The steady-state equations that have to be solved are

$$0 = b_1 - (k_0 + k_1 x_6) x_1, \tag{9.37}$$

$$0 = k_1 x_6 x_1 - k_2 x_2, \tag{9.38}$$

$$0 = k_2 x_2 - k_3 x_3, \tag{9.39}$$

$$0 = k_3 x_3 - k_4 x_4, \tag{9.40}$$

$$0 = k_4 x_4 - k_5 x_5 - (k_6 x_5 x_6 - k_{-6} x_7), \tag{9.41}$$

$$0 = k_6 x_5 x_6 - k_{-6} x_7, \tag{9.42}$$

$$0 = x_6 + x_7 - e_t. \tag{9.43}$$

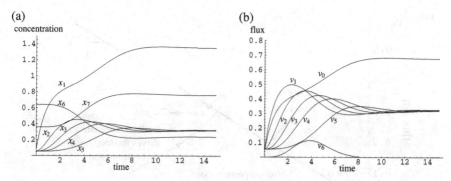

Figure 9.8 The time profiles for the concentrations and fluxes involved in the simple feedback control loop. The parameter values used are $k_0 = 0.5$, $k_1 = 1$, $k_2 = 1$, $k_3 = 1$, $k_4 = 1$, $k_5 = 1$, $k_6 = 10$, $k_{-6} = 1$, $e_t = 1.0$. Just prior to time zero, the input rate is $b_1 = 0.1$ and the feedback loop is at steady state, where $x_1 = 0.088$, $x_2 = x_3 = x_4 = x_5 = 0.0562$, $x_6 = 0.640$, and $x_7 = 0.360$. The input rate is changed to $b_1 = 1.0$ at time zero and the dynamic response is simulated. (a) The concentrations as a function of time. (b) The reaction fluxes as a function of time.

These equations can be combined to give a quadratic equation,

$$y^2 + ay - b = 0, \tag{9.44}$$

where

$$y = k_2 x_2, \quad a = k_5 \left(\frac{k_{-6}}{k_6}\right)\left(1 + \frac{k_1 e_t}{k_0}\right), \quad b = k_5 \left(\frac{k_{-6}}{k_6}\right)\left(\frac{k_1 e_t}{k_0}\right) b_1, \tag{9.45}$$

that has one positive root.

Dynamic simulations After the numerical values for the parameters have been specified, we can simulate these equations numerically (Figures 9.8 and 9.9). In this simulation, the feedback loop is started out at steady state before time zero. At time zero, the input rate b_1 is increased 10-fold: it shifts from 0.1 to 1.0. The increased input rate leads to an immediate increase in x_1. Then, the intermediates in the reaction chain (x_2, x_3, x_4, x_5) rise serially over time; see Figure 9.8a. As x_5 starts building up, it begins to sequester the enzyme (x_6) in its inactive form (x_7) through activation of v_6; see Figure 9.8b. As the free enzyme concentration drops, the influx to the reaction chain (v_1) and the inhibitory flux (v_6) drops towards its steady state. Note that the concentrations and fluxes overshoot their eventual steady state.

These complex dynamics can be further examined and simplified using dynamic phase portraits. The input flux (v_1) into the reaction sequence can be graphed versus the output rate; see Figure 9.9a. At time zero, the input and output fluxes are equal, $v_1 = v_5$. With the immediate increase in x_1, the input flux increases parallel to the x-axis. Then v_5 starts to increase,

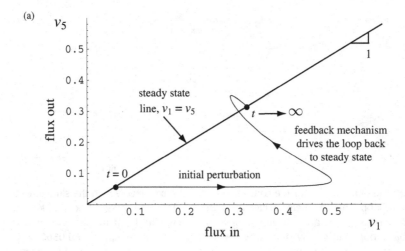

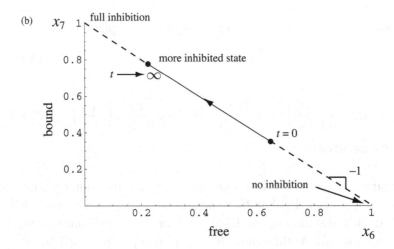

Figure 9.9 Dynamic phase portraits for the simple feedback loop for the same conditions as in Figure 9.8. (a) Flux into (v_1) and out of (v_5) in the reaction sequence. (b) The phase portrait of x_6 and x_7 showing that the two move on the conservation line $x_6 + x_7 = 1$.

leading to a drop in x_6, and thus v_1 drops. The trajectory overshoots the 45° line (i.e., where $v_1 = v_5$), turns around, and eventually comes to a rest on the 45° line. The mass balance between x_6 and x_7 can be seen in Figure 9.9b. The dynamic phase portrait shows the motion of the two on a line with a slope of -1.

For this numerical example, the feedback control is not working very well, as v_1 does not return close to its original value, and the system is not very 'robust' with respect to changes in b_1. Once the equations have been set up, simulations can be repeated from different initial conditions and parameter values. For example, the user can adjust the inhibition by changing the values of k_6 and k_{-6}. Specifically, if we set $k_6 = 0$, then there is no inhibition. Conversely, if k_6 and k_6/k_{-6} are high, then the inhibition is rapid and tight.

Conserved pools and their state This example introduces an important concept of an enzyme pool size and its state. In this case, the total enzyme $(e_t = x_6 + x_7)$ is a constant. The fraction

$$f_1 = \frac{x_7}{x_6 + x_7} \tag{9.46}$$

is the fraction of the total enzyme that is inhibited and prevented from catalyzing the reaction. This sequestering of catalytic activity is essentially how the feedback works. At time zero, $f_1 = 0.36$; thus, about a third of the enzyme is in a catalytically inactive state. When the b_1 flux is increased, the concentration of x_1 builds and increases the flux (v_1) into the reaction chain. The subsequent buildup of the end product (x_5) leads to an adjustment of the inhibited fraction of the enzyme to $f_1 = 0.74$ and, thus, reduces the input flux into the reaction chain.

9.6 Increasing network complexity

Building on a scaffold The reaction network studied in the previous section represents the basic framework for feedback regulation of a biosynthetic pathway. This schema operates in a cellular environment that has additional regulatory features. These can be built on top of this basic structure.

We will consider two additional regulatory mechanisms. First, we look at the regulation of protein synthesis and more elaborate and realistic schemas for the inhibition of the first enzyme in the pathway. Second, we consider more and more realistic mechanisms for x_5 binding to the regulated enzyme. The two can be combined.

Such additions take into account more processes and make the models more realistic. However, the number of parameters grows and the issue of getting accurate numerical values for them becomes more challenging.

9.6.1 Regulation of protein synthesis
In the previous example, x_5 feedback inhibits its own synthesis by transforming the first enzyme in the reaction chain into an inactive form. In many organisms, end products like x_5 can also feedback regulate the

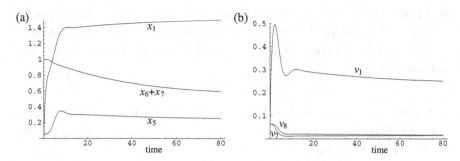

Figure 9.10 The time profiles for concentrations and fluxes involved in the simple feedback control loop with feedback control of protein synthesis. The parameter values used are $k_7 = 0.70$, $K_7 = 178$, and $k_8 = 0.10$. Other parameter values are as in Figure 9.8. (a) Key concentrations as a function of time. (b) Key reaction fluxes as a function of time.

synthesis rate of the enzyme itself and thus regulate the total amount of enzyme present. Thus, e_t is no longer a constant, but is a dynamic variable.

Additions to the dynamic equations   This regulation of protein synthesis can be simulated by adding a synthesis and degradation rate in the dynamic equation for the enzyme, x_6:

$$\frac{dx_6}{dt} = -k_6 x_5 x_6 + k_{-6} x_7 + v_7 - v_8,$$ (9.47)

where we can use an inhibition rate of the Hill form with $v = 1$:

$$v_7 = \frac{k_7}{1 + K_7 x_5}$$ (9.48)

and a first-order turnover

$$v_8 = k_8 x_6$$ (9.49)

for the enzyme. Note that these equations assume that the inhibited form x_7 is stable.

The addition of the regulation of protein synthesis adds two new rate laws with three new parameters. Such addition may call on difficult experiments to determine the numerical values of k_7, K_7, and k_8. Such expansion in the number of parameters and experimentation shows the challenge with scaling kinetic models and increasing their scope.

Dynamic simulation   The effect of this additional process can be simulated. If we pick $k_7 = 0.70$, $K_7 = 178$, and $k_8 = 0.10$, then the network will have the same initial state as shown in Figure 9.8. The stability of the enzyme ($k_8 = 0.10$) introduces a slower secondary response of this loop. During this secondary response, the total enzyme concentration drops from 1.0 to a value of 0.59 (Figure 9.10a). During this slow response, the concentration of x_1 rises due to a lower flux into the reaction chain (v_1) and the end product (x_5) drops slightly (Figure 9.10b).

Slowly changing pool sizes   These changes can also be traced out using dynamic phase portraits; see Figure 9.11. Compared with the response without the regulation of protein synthesis (Figure 9.9a), the input and output rates from the reaction chain drop slowly (Figure 9.11a). This motion is approximately along the 45° line. The phase portrait of the two forms of the enzyme (x_6 and x_7) initially follows the straight line $x_6 + x_7 = 1$, as without the control of enzyme synthesis (Figure 9.9b), but then has a secondary slow response where the total amount of the enzyme drops from 1.0 to about 0.59. Thus, the control of enzyme synthesis drops the total amount of the enzyme by about 40%, leading to a lower flux through v_1. Notice, however, that the buildup of x_1 tempers the drop in v_1. The flux into the reaction chain starts at 0.06, rapidly increases to above 0.5, is pulled back to about 0.3 by enzyme inhibition, and then drops slowly to about 0.25 with regulation of protein synthesis. Thus, even this bi-level regulation is not able to reject the disturbance caused by a 10-fold increase in b_1.

9.6.2 Tight regulation of enzyme activity

The feedback inhibition of enzyme activity from the previous section is not very effective. As one can see in Figure 9.9a, the feedback loop does not return the system to a state that is close to the initial state. This means that the feedback loop cannot "reject" the disturbance in the input flux well.

More realistic mechanism for x_5 binding   Regulatory enzymes have a more complex mechanism than simply having an active and inactive state, as denoted by x_6 and x_7 in Section 9.5. Regulatory enzymes often have a series of binding sites for inhibitory molecules. One mechanism for such serial binding is the symmetry model, described in Section 5.5. The reaction mechanisms (using the same compound names as in Section 9.5) are, for a dimer,

$$x_6 + x_5 \underset{k_{-6}}{\overset{2k_6}{\rightleftharpoons}} x_7, \quad x_7 + x_5 \underset{2k_{-6}}{\overset{k_6}{\rightleftharpoons}} x_8, \tag{9.50}$$

where x_7 has one ligand bound and x_8 has two, and for a tetramer,

$$x_6 + x_5 \underset{k_{-6}}{\overset{4k_6}{\rightleftharpoons}} x_7, \quad x_7 + x_5 \underset{2k_{-6}}{\overset{3k_6}{\rightleftharpoons}} x_8,$$

$$x_8 + x_5 \underset{3k_{-6}}{\overset{2k_6}{\rightleftharpoons}} x_9, \quad x_9 + x_5 \underset{4k_{-6}}{\overset{k_6}{\rightleftharpoons}} x_{10}, \tag{9.51}$$

where x_9 and x_{10} have three and four ligands bound respectively.

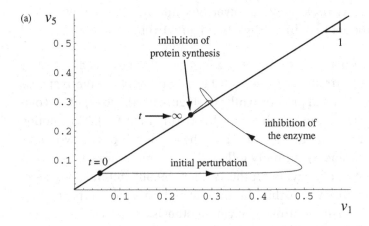

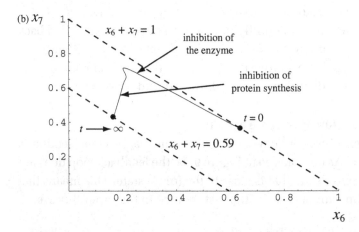

Figure 9.11 Dynamic phase portraits for the simple control loop for the same conditions as in Figure 9.10. (a) Flux into (v_1) and out of (v_5) in the reaction sequence. (b) The phase portrait of x_6 and x_7 showing that the two move initially on the conservation line $x_6 + x_7 = 1$, followed by a slower response where the total enzyme drops to 0.59.

Dynamic simulation   These two mechanisms can be separately analyzed in terms of their efficacy to improve the same disturbance rejection as simulated in Section 9.5. The results are plotted in Figure 9.12, which is analogous to Figure 9.9. The return and balancing of the input (v_1) and output (v_5) fluxes from the biosynthetic pathway return much closer to the original state of the system. The tetramer is quite effective for rejecting the disturbance.

Pool sizes and their states   The fraction of the enzyme that is in the inhibited form can be computed for the dimer as

$$f_2 = \frac{x_7 + 2x_8}{2(x_7 + x_8)} \tag{9.52}$$

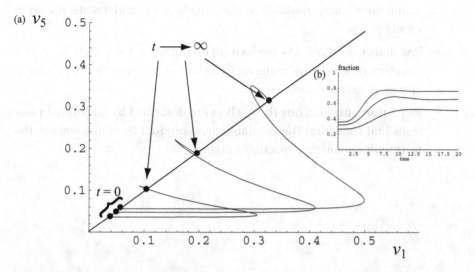

Figure 9.12 Dynamic phase portraits for the simple feedback loop for the same conditions as in Figure 9.10, except with dimeric and tetrameric forms of the inhibition mechanism. (a) Flux into (v_1) and out of (v_5) in the reaction sequence. (b) The fraction of the enzyme that is in an inhibited form as a function of time.

and for the tetramer as

$$f_4 = \frac{x_7 + 2x_8 + 3x_9 + 4x_{10}}{4(x_7 + x_8 + x_9 + x_{10})}. \tag{9.53}$$

In both cases, these fractions are the ratio of the occupied sites to the total number of binding sites for x_5. The increased number of binding sites for x_5 on the enzyme (x_6) increases the efficacy of the regulatory mechanisms on the ability of the system to reject the disturbance imposed on it.

9.7 Summary

➤ The activities of gene products are often directly regulated by many different genetic and biochemical mechanisms.

➤ From a systems standpoint, regulation can be described by (i) its bias, (ii) the concentration range over which the regulatory molecule is active, and (iii) its strength, or how sensitive the flux is to changes in the concentration of the regulator.

➤ The "distance" in the network between the site of regulation and the formation of the regulator is an important consideration.

➤ In general, local signals that support the natural mass action trend in a network are "stabilizing" and local signals that counter the mass action trend may destabilize the steady state and create multiple steady states.

➤ Regulatory mechanisms can be built on top of the basic stoichiometric structure of a network being analyzed and its description by elementary mass action kinetics.

➤ Regulatory mechanisms themselves are described by additional reactions that transform the regulated gene product from one state to the next with elementary reaction kinetics.

Metabolism

The procedures laid out in Part II lead to the formulation of a MASS model-building process. The process is a step-by-step increase in the scope and coverage of a model by systematically increasing the size of the stoichiometric matrix. Mass action kinetics are assumed to represent the dynamics of the reaction rates. Typical numerical values of the concentrations and fluxes are then used to get PERCs. This leads to a condition-dependent kinetic model. The unregulated version of MASS models was covered in Chapter 8 and the incorporation of regulation was outlined in Chapter 9.

In the third part of the book we develop a series of MASS models of increasing scope for a realistic biochemical network using metabolism in the human red blood cell as an example. We begin with the classical glycolytic pathway. We then couple the pentose pathway to glycolysis to study the resulting effects. Finally, we add nucleotide metabolism to form an integrated network.

This material shows how kinetic models are built, analyzed, and scaled up to physiologically meaningful sizes based on purely network structure and available concentration and fluxomic data. Part IV will then show how the activities of protein can be incorporated into this structure.

## CHAPTER 10

Glycolysis

Glycolysis is a central metabolic pathway. In this chapter, we formulate a MASS model of the glycolytic pathway. In doing so, we detail the process that goes into formulating, characterizing, and validating a simulation model of a single pathway. First, we define the pathway, or, more accurately stated, the *system* that is to be studied and characterized through dynamic simulation. Such a definition includes the internal reactions, the systems boundary, and the exchange fluxes. The stoichiometric matrix is then constructed and quality controlled, and its properties studied. The contents of the null spaces give us information about the pathway and pool structure of the defined system. The steady state is deduced through the specification of a minimum number of fluxes. The steady-state concentrations are then specified and the mass action kinetic constants are evaluated. Finally, the functional pooling structure and corresponding property ratios are formulated using biochemical rationale. The dynamic responses of glycolysis as a system can then be simulated and interpreted through the pooling structure. As in Chapter 8, we focus on the response to an increase in the rate of ATP utilization. With the formulated glycolytic MASS model, the reader can simulate and study responses to other perturbations.

10.1 Glycolysis as a system

Defining the system The glycolytic pathway degrades glucose (a six-carbon compound) to form pyruvate or lactate (three-carbon compounds) as end products. During this degradation process, the glycolytic pathway builds redox potential in the form of NADH and high-energy phosphate bonds in the form of ATP via substrate-level phosphorylation. Glycolysis also assimilates an inorganic phosphate group that is converted into

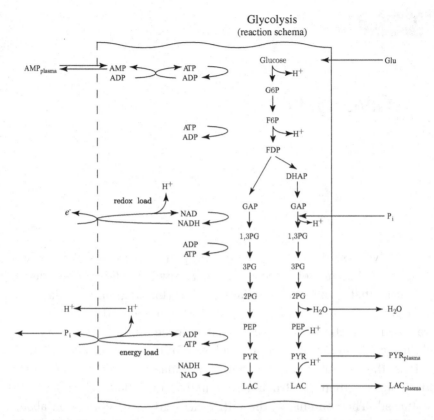

Figure 10.1 Glycolysis: the reaction schema, cofactor interactions, and environmental exchanges.

a high-energy bond and then hydrolyzed in the ATP use reaction (or the "load" reaction).

Glycolysis thus has interactions with three key cofactor or carrier moieties (ATP, NADH, P_i). These three key attributes of the glycolytic pathway can be examined from a biochemical and metabolic physiological standpoint (see Section 10.5) and will influence how glycolysis interacts with other cellular functions. Glycolysis as a system is shown in Figure 10.1. In this chapter we focus only on the fundamental stoichiometric characteristics of the glycolytic system, but in subsequent chapters we will systematically expand the scope of the system being studied by including the regulatory enzymes.

The nodes (compounds) and links (reactions)   In its simplest form, glycolysis has 12 primary metabolites, five cofactor molecules (ATP, ADP, AMP, NAD, NADH), inorganic phosphate (P_i), protons (H^+), and water (H_2O). The system thus has 20 compounds; see Table 10.1. The links

Table 10.1 The compounds in the glycolytic system, their abbreviations, and representative steady-state concentrations. These compounds can be divided into three groups: (i) the glycolytic intermediates (1–12), (ii) the cofactors (13–17), and (iii) inorganic components (18–20). The concentration of water is arbitrary as it will not appear in any reaction rate laws

#	Abbreviation	Intermediates and cofactors	Concentration (mM)
1	Gluc	Glucose	1.0
2	G6P	Glucose 6-phosphate	0.0486
3	F6P	Fructose 6-phosphate	0.0198
4	FDP	Fructose 1,6-diphosphate	0.0146
5	DHAP	Dihydroxyacetone phosphate	0.16
6	GAP	Glyceraldehyde 3-phosphate	0.00728
7	DPG13	1,3-Diphosphoglycerate	0.00024
8	PG3	3-Phosphoglycerate	0.0773
9	PG2	2-Phosphoglycerate	0.0113
10	PEP	Phosphoenolpyruvate	0.017
11	PYR	Pyruvate	0.0603
12	LAC	Lactate	1.36
13	NAD	Nicotinamide adenine dinucleotide (oxidized)	0.0589
14	NADH	Nicotinamide adenine dinucleotide (reduced)	0.0301
15	AMP	Adenosine mono-phosphate	0.0867
16	ADP	Adenosine di-phosphate	0.29
17	ATP	Adenosine tri-phosphate	1.6
18	P_i	Inorganic phosphate	2.5
19	H^+	Proton	10^{-4}
20	H_2O	Water	arbitrary

formed between these compounds are the glycolytic reactions. There are 21 reactions, including all the transport reactions in and out of the system. The reactions are summarized in Table 10.2.

10.2 The stoichiometric matrix

The stoichiometric matrix **S** can be formulated for the glycolytic system (see Table 10.3). Its dimensions are 20×21, representing the 20 metabolites and the 21 fluxes given in Tables 10.1 and 10.5 respectively. The rank of this stoichiometric matrix is 18, leaving a three-dimensional null space (i.e., $21 - 18$) and a two-dimensional left null space (i.e., $20 - 18$). The

Table 10.2 The glycolytic enzymes and transporters, their abbreviations, and chemical reactions. These reactions can be grouped into several categories: (i) the glycolytic reactions (1–11), (ii) adenosine phosphate metabolism (12,13), (iii) primary exports (14,15), (iv) the load functions (16,17), (v) the primary inputs, here fixed (18,19), and (vi) the water and proton exchanges

#	Abbrev.	Enzymes/transporter/load	Elementally balanced reaction
1	HK	Hexokinase	$Gluc + ATP \rightarrow G6P + ADP + H^+$
2	PGI	Glucose 6-phosphate isomerase	$G6P \leftrightarrow F6P$
3	PFK	Phosphofructokinase	$F6P + ATP \rightarrow FDP + ADP + H^+$
4	TPI	Triose-phosphate isomerase	$DHAP \leftrightarrow GAP$
5	ALD	Fructose 1,6-diphosphate aldolase	$FDP \leftrightarrow DHAP + GAP$
6	GAPDH	Glyceraldehyde 3-phosphate dehydrogenase	$GAP + NAD + P_i \leftrightarrow DPG13 + NADH + H^+$
7	PGK	Phosphoglycerate kinase	$DPG13 + ADP \leftrightarrow PG3 + ATP$
8	PGLM	Phosphoglycerate mutase	$PG3 \leftrightarrow PG2$
9	ENO	Enolase	$PG2 \leftrightarrow PEP + H_2O$
10	PK	Pyruvate kinase	$PEP + ADP + H+ \rightarrow PYR + ATP$
11	LDH	Lactate dehydrogenase	$PYR + NADH + H+ \leftrightarrow LAC + NAD$
12	AMP	AMP export	$AMP \rightarrow$
13	APK	Adenylate kinase	$2ADP \leftrightarrow AMP + ATP$
14	PYR	Pyruvate exchange	$PYR \leftrightarrow$
15	LAC	Lactate exchange	$LAC \leftrightarrow$
16	ATP	ATP hydrolysis	$ATP + H_2O \rightarrow ADP + P_i + H^+$
17	NADH	NADH oxidation	$NADH \rightarrow NAD + H^+$
18	GLU_{in}	Glucose import	$\rightarrow Gluc$
19	AMP_{in}	AMP import	$\rightarrow AMP$
20	H	Proton exchange	$H^+ \leftrightarrow$
21	H_2O	Water exchange	$H_2O \leftrightarrow$

stoichiometric matrix has many attributes and properties; recall Table 1.3. The ones from a modeling and systems standpoint will now be discussed.

Elemental balancing   The stoichiometric matrix needs to be quality controlled to make sure that the chemical equations are mass balanced. The elemental compositions of the compounds in the glycolytic system are given in Table 10.4. This table is the elemental matrix **E** for this system. We can multiply **ES** to quality control the reconstructed network for elemental balancing properties of the reactions (i.e., verify that **ES = 0**). The results are shown in Table 10.3. All the internal reactions are elementally

Table 10.3 The stoichiometric matrix for the glycolytic system in Figure 10.1. The matrix is partitioned to show the cofactors separate from the glycolytic intermediates and to separate the exchange reactions and cofactor loads. The last column has the connectivities ρ_I for a compound, and the last row has the participation number π_J for a reaction. The second block in the table is the product **ES** to evaluate elemental balancing status of the reactions. All exchange reactions have a participation number of unity and are thus elementally balanced. The last block in the table has the pathway vectors for glycolysis. These vectors are shown graphically in Figure 10.2. The last line shows the steady-state fluxes (in mM/h) computed from Eq. (10.3)

	Glycolytic reactions											AMP metabolism		Primary export		Cofactors		Primary inputs		Inorganic		
	v_{hk}	v_{pgi}	v_{pfk}	v_{tpi}	v_{ald}	v_{gapdh}	v_{pgk}	v_{pgm}	v_{eno}	v_{pk}	v_{ldh}	v_{amp}	v_{apk}	v_{pyr}	v_{lac}	v_{atp}	v_{nadh}	v_{gluin}	v_{ampin}	v_{H^+}	v_{H_2O}	ρ_I
Glu	−1	0	0	0	0	0	0	0	0	0	0	0	0	0	0	0	0	1	0	0	0	2
G6P	1	−1	0	0	0	0	0	0	0	0	0	0	0	0	0	0	0	0	0	0	0	2
F6P	0	1	−1	0	0	0	0	0	0	0	0	0	0	0	0	0	0	0	0	0	0	2
FBP	0	0	1	0	−1	0	0	0	0	0	0	0	0	0	0	0	0	0	0	0	0	2
DHAP	0	0	0	−1	1	0	0	0	0	0	0	0	0	0	0	0	0	0	0	0	0	2
GAP	0	0	0	1	1	−1	0	0	0	0	0	0	0	0	0	0	0	0	0	0	0	3
PG13	0	0	0	0	0	1	−1	0	0	0	0	0	0	0	0	0	0	0	0	0	0	2
PG3	0	0	0	0	0	0	1	−1	0	0	0	0	0	0	0	0	0	0	0	0	0	2
PG2	0	0	0	0	0	0	0	1	−1	0	0	0	0	0	0	0	0	0	0	0	0	2
PEP	0	0	0	0	0	0	0	0	1	−1	0	0	0	0	0	0	0	0	0	0	0	2
PYR	0	0	0	0	0	0	0	0	0	1	−1	0	0	−1	0	0	0	0	0	0	0	3
LAC	0	0	0	0	0	0	0	0	0	0	1	0	0	0	−1	0	0	0	0	0	0	2
NAD	0	0	0	0	0	−1	0	0	0	0	1	0	0	0	0	0	1	0	0	0	0	3
NADH	0	0	0	0	0	1	0	0	0	0	−1	0	0	0	0	0	−1	0	0	0	0	3
AMP	0	0	0	0	0	0	0	0	0	0	0	−1	1	0	0	0	0	0	1	0	0	3

(continued)

Table 10.3 (continued).

	Glycolytic reactions											AMP metabolism		Primary export		Cofactors		Primary inputs		Inorganic		
	v_{hk}	v_{pgi}	v_{pfk}	v_{tpi}	v_{ald}	v_{gapdh}	v_{pgk}	v_{pglm}	v_{eno}	v_{pk}	v_{ldh}	v_{amp}	v_{apk}	v_{pyr}	v_{lac}	v_{atp}	v_{nadh}	v_{gluin}	v_{ampin}	v_{H^+}	v_{H_2O}	z_j
ADP	1	0	1	0	0	0	−1	0	0	−1	0	0	−2	0	0	1	0	0	0	0	0	6
ATP	−1	0	−1	0	0	0	1	0	0	1	0	0	1	0	0	−1	0	0	0	0	0	6
P_i	0	0	0	0	0	−1	0	0	0	−1	0	0	0	0	0	1	0	0	0	0	0	2
H^+	1	0	1	0	0	1	0	0	1	0	−1	0	0	0	0	−1	1	0	0	−1	0	8
H_2O	0	0	0	0	0	0	0	0	1	0	0	0	0	0	0	−1	0	0	0	0	−1	3
π_j	5	2	5	2	3	5	4	2	3	5	5	1	3	1	1	5	3	1	1	1	1	
C	0	0	0	0	0	0	0	0	0	0	0	−10	0	−3	−3	0	0	6	10	0	0	
H	0	0	0	0	0	0	0	0	0	0	0	−13	0	−3	−5	0	0	12	13	−1	−2	
O	0	0	0	0	0	0	0	0	0	0	0	−7	0	−3	−3	0	0	6	7	0	−1	
P	0	0	0	0	0	0	0	0	0	0	0	−1	0	0	0	0	0	0	1	0	0	
N	0	0	0	0	0	0	0	0	0	0	0	−5	0	0	0	0	0	0	5	0	0	
S	0	0	0	0	0	0	0	0	0	0	0	0	0	0	0	0	0	0	0	0	0	
NAD	0	0	0	0	0	0	0	0	0	0	0	0	0	0	0	0	0	0	0	0	0	
$\mathbf{p_1}$	1	1	1	1	1	2	2	2	2	2	2	0	0	0	2	2	0	1	0	2	0	0
$\mathbf{p_2}$	0	0	0	0	0	0	0	0	0	0	−1	0	0	1	−1	0	1	0	0	2	0	0
$\mathbf{p_3}$	0	0	0	0	0	0	0	0	0	0	0	1	0	0	0	0	0	0	1	0	0	0
$\mathbf{v_{stst}}$	1.12	1.12	1.12	1.12	1.12	2.24	2.24	2.24	2.24	2.24	2.016	0.014	0	0.224	2.016	2.24	0.224	1.12	0.014	2.69	0	

Table 10.4 The elemental composition of the glycolytic intermediates, see Table 10.1 for details. NAD is treated as one moiety. This table represents the matrix **E**

	Glu	G6P	F6P	FBP	DHAP	GAP	PG13	PG3	PG2	PE	PYR	LAC	NAD	NADH	AMP	ADP	ATP	P_i	H^+	H_2O
C	6	6	6	6	3	3	3	3	3	3	3	3	0	0	10	10	10	0	0	0
H	12	11	11	10	5	5	4	4	4	2	3	5	0	1	13	13	13	1	1	2
O	6	9	9	12	6	6	10	7	7	6	3	3	0	0	7	10	13	4	0	1
P	0	1	1	2	1	1	2	1	1	1	0	0	0	0	1	2	3	1	0	0
N	0	0	0	0	0	0	0	0	0	0	0	0	0	0	5	5	5	0	0	0
S	0	0	0	0	0	0	0	0	0	0	0	0	0	0	0	0	0	0	0	0
NAD	0	0	0	0	0	0	0	0	0	0	0	0	1	1	0	0	0	0	0	0

balanced. The exchange reactions are not elementally balanced as they represent net addition or removal from the system as defined.

Charge balancing In this chapter, we will treat the compounds as being uncharged. This assumption is not physiologically accurate, but it will not affect our dynamic simulations. If significant changes in pH are to be considered, then the charged state of the molecules needs to be established. A row can be added into **E** representing the charges of the molecules. Then, charge balance is ensured by making sure that $\mathbf{ES} = \mathbf{0}$. Charge balancing the whole system for transporters can be difficult, as some of the transport systems, co-transport ions, and some ions can cross the membrane by themselves. Overall, the system has to be charge neutral. Accounting for full charge balances and the volume of a system can be quite mathematically involved; see [59, 61].

Topological properties The simplest topological properties of the stoichiometric matrix relate to the number of nonzero elements in rows and columns. The more advanced topological properties are associated with the null spaces and their basis vectors and are quantities of systems biology.

The number of reactions in which a compound participates, ρ_i, is the *connectivity*. It is formally defined as the total number of nonzero elements in a row [89]. Most of the compounds are formed by one reaction and degraded by another, thus having a connectivity of two. The proton participates in eight reactions, while ATP and ADP participate in six, making them the most connected compounds in the system.

The number of compounds that participate in a reaction, π_j, is called the *participation* number. It is formally defined as the total number of

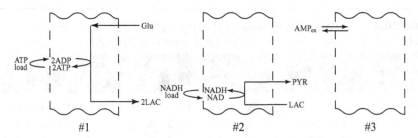

Figure 10.2 The pathway vectors for glycolysis. They span all possible steady-state solutions.

nonzero elements in a column [89]. Exchange reactions have a participation of unity and are thus elementally imbalanced. Some of the kinases and dehydrogenases have the highest participation number, 5.

The pathway structure: basis for the null space   The null space is spanned by three vectors, p_1, p_2, p_3, in Table 10.3, that have pathway interpretations (Chapter 9 of [89]). These three pathways are shown in Figure 10.2, and they are interpreted as follows:

1. The first pathway is a redox-balanced use of glycolysis to produce two high-energy bonds per glucose metabolized. A glucose enters and two lactates leave. There is no net production of NADH, but there is a net production of 2ATP that is balanced by the ATP load reaction.
2. The second pathway describes how a redox load on NADH is balanced by the uptake of lactate to produce pyruvate via LDH and generate the NADH that is used in the NADH load reaction. When this pathway is added to the first one, it leads to a reduced lactate and pyruvate secretion and the net production to NADH that meets the load imposed.
3. The third pathway is simply AMP entering and leaving the system. These rates have to be balanced in a steady state. This pathway will determine the total amount ($A_{tot} = ATP + ADP + AMP$) of the adenosine phosphates in the system. In Chapter 12, we detail the synthesis of AMP through the addition of the pentose and salvage pathways.

The non-negative addition of these three pathways would give the steady-state flux distribution, as discussed in Section 10.3. Although these three vectors are a basis for the null space, they are not a unique basis. There are other choices of linear basis vectors. Alternatively, the so-called convex basis vectors do form a unique basis [89, 94].

Elemental balancing of the pathway vectors Each of the pathway vectors needs to be elementally balanced, which means the elements coming into and leaving a pathway through the transporters need to be balanced. Thus, we must have $\mathbf{ESp}_i = 0$ for each pathway. This condition is satisfied for these three pathways. One can readily ascertain this balance by adding up the columns in Table 10.3 that correspond to the exchange reactions in a pathway.

The time-invariant pools: the basis for the left null space The stoichiometric matrix has a left null space of dimension 2, meaning the glycolytic system as defined has two time-invariant pools.

1. The first time-invariant is the total amount of phosphates on all the compounds in the network (looking ahead, we put this as pool number 10 and second to last line in Table 10.7). Mathematically it is

$$p_{10} = \text{G6P} + \text{F6P} + 2\text{FDP} + \text{DHAP} + \text{GAP} + 2\text{DPG13} + \text{PG3} + \text{PG2}$$

$$+ \text{PEP} + \text{ADP} + 2\text{ATP} + \text{P}_i$$

$$= \text{P}_{\text{tot}}. \tag{10.1}$$

   Notice that the phosphate on AMP is not counted, as this phosphate group enters and leaves the system as a part of the AMP moiety.

2. The second time-invariant is the total NADH balance. This cofactor never leaves or enters the system as defined; Figure 10.1. It appears as the last line in Table 10.7, and is

$$p_{11} = \text{NADH} + \text{NAD} = \text{N}_{\text{tot}} \tag{10.2}$$

Similarly, the linear basis for the left null space is not unique, but a convex basis is [32, 89].

10.3 Defining the steady state

So far, we have defined the topological structure of the glycolytic system and used it to perform quality controls related to chemistry. We have also determined topological properties, namely the pathways and invariant pools, by looking at the basis vectors for the null spaces. We will now introduce (1) flux data to determine the steady state flux map and (2) steady-state metabolite concentrations that will lead to the computation of the PERCs. Note that the PERCs are condition dependent, as they depend on the flux and concentration measurements. Numerical values

for the concentrations and exchange fluxes need to be obtained experimentally. In principle, comprehensive fluxomic and metabolomic data
sets should enable this process to be performed at a larger scale.

Computing the steady-state flux map   Flux and concentration data for
glycolysis are available for several organisms and cell types. Here, we will
use data from the classical human red blood cell metabolic model [42, 55,
58, 60, 61] as an example. The null space of S is three-dimensional. Thus,
the specification of three independent fluxes allows the determination of
the unique steady-state flux state. The steady-state flux distribution is a
combination of the three pathway vectors.

The uptake rate of the red blood cell of glucose is about 1.12 mM/h. This
number specifies the length of the first pathway vector, p_1, in the steady-
state solution, v_{stst}. Based on experimental data, the steady-state load on
NADH is set at 20% of the glucose uptake rate, or $0.2 \times 1.12 = 0.224$.
This number specifies the length of the second pathway vector, p_2, in the
steady-state flux vector. The input of AMP is measured to be 0.014 mM/h.
This number specifies the length of the third pathway vector, p_3, in the
steady-state flux vector. Thus, we can compute the steady-state flux vector
as

$$v_{stst} = 1.12 p_1 + 0.224 p_2 + 0.014 p_3. \tag{10.3}$$

The steady-state flux map is computed and shown in the last row of
Table 10.3. Note that there is a net production of a proton, leading to
acidification of the surrounding medium, but no net production of water.

Computing the rate constants   Glycolysis contains both reversible and
effectively irreversible reactions. The numerical values for their equilibrium constants are given in Table 10.5. The approximate steady-state values of the metabolites are in Table 10.1. The mass action ratios (recall
Eq. (2.11)) can be computed from these steady-state concentrations and are
all smaller than the corresponding equilibrium constants (i.e., $\Gamma < K_{eq}$),
and thus the reactions are proceeding in the forward direction (Table 10.5).

We can compute the forward rate constant for reaction i from

$$k_i = \frac{\text{flux through reaction}}{(\Pi_i \text{ reactants}_i - \Pi_i \text{ products}_i / K_{eq})}. \tag{10.4}$$

These estimates for the numerical values for the PERCs are also contained
in Table 10.5. These numerical values, along with the elementary form
of the rate laws, complete the definition of the dynamic mass balances
that can now be simulated. The steady state is specified in Table 10.6.
Clearly, there are practical limitations for the computation of the PERCs

Table 10.5 Glycolytic enzymes, AMP metabolism, loads, transport rates, and their abbreviations. In addition, the glucose influx rate is set at 1.12 mM/h and the AMP influx rate is set at 0.014 mM/h based on data; see [58]. For irreversible reactions, the numerical value for the equilibrium constants is ∞, which, for practical reasons, has to be set to a finite value, here at 10^6

	Abbrev.	Enzymes/transporter/load	Equilibrium constant	Value	PERC	Value	Mass action ratio (Γ)	Γ/K_{eq}
1	HK	Hexokinase	K_{HK}	850	k_{HK}	0.70	0.009	0
2	PGI	Phosphoglucoisomerase	K_{PGI}	0.41	k_{PGI}	3464.4	0.407	0.994
3	PFK	Phosphofructokinase	K_{PFK}	310	k_{PFK}	35.37	0.134	0
4	ALD	Aldolase	K_{ALD}	0.082	k_{ALD}	2834.57	0.08	0.973
5	TPI	Triose phosphate isomerase	K_{TPI}	0.0571	k_{TPI}	34.36	0.046	0.796
6	GAPDH	GAP dehydrogenase	K_{GADPH}	0.018	k_{GADPH}	3376.75	0.007	0.381
7	PGK	Phosphoglycerate kinase	K_{PGK}	1800	k_{PGK}	1.274×10^6	1755.07	0.975
8	PGM	Phosphoglyceromutase	K_{PGM}	0.147	k_{PGM}	4869.57	0.146	0.994
9	ENO	Enolase	K_{ENO}	1.695	k_{ENO}	1763.78	1.504	0.888
10	PK	Pyruvate kinase	K_{PK}	363 000	k_{PK}	454.386	19.57	0
11	LDH	Lactate dehydrogenase	K_{LDH}	26300	k_{LDH}	1112.57	44.133	0.002
12	AMP	Adenosine monophosphate removal	K_{AMP}	10^6	k_{AMP}	0.161	0.001	0
13	ApK	Adenylate kinase	K_{ApK}	1.65	k_{ApK}	1	1.65	1
14	PYR$_{ex}$	Pyruvate export	K_{PYR}	1	k_{PYR}	744.186	0.995	0.995
15	LAC$_{ex}$	Lactate export	K_{LAC}	1	k_{LAC}	5.60	0.735	0.735
16	ATP$_{load}$	Energy load	K_{ATP}	10^6	k_{ATP}	1.40	0.453	0
17	NADH$_{load}$	Redox load	K_{NADH}	10^6	k_{NADH}	7.441	1.957	0
18	Glu$_{in}$	Glucose in (fixed value of 1.12 mM/h)						
19	AMP$_{in}$	AMP in (fixed value of 0.014 mM/h)						
20	H$^+$	Freely exchanging proton	K_{H^+}	1				
21	H$_2$O	Freely exchanging water	K_{H_2O}	1				

Table 10.6 Numerical values for the concentrations (mM) and fluxes (mM/h) at the beginning and end of the dynamic simulation. The concentration of water is arbitrarily set at 1.0. At $t = 0$ the v_{ATP} flux changes from 2.24 to 3.36, misbalancing the flux map. The flux map returns to its original state as time goes to infinity. Notice here that $t = 50$ is effectively not infinity, as some of the fluxes, such as v_{ApK}, have not fully reached the original value. Also note that unlike the fluxes, the concentrations will reach a different steady state

Compound	$t = 0^+$	$t = 50$	Flux	$t = 0^+$	$t = 50$
Glu	1.0	1.50	v_{HK}	1.12	1.12
G6P	0.0486	0.0727	v_{PGI}	1.12	1.12
F6P	0.0198	0.0297	v_{PFK}	1.12	1.12
FBP	0.0146	0.00741	v_{TPI}	1.12	1.12
DHAP	0.16	0.118	v_{ALD}	1.12	1.12
GAP	0.00728	0.00488	v_{GAPDH}	2.24	2.24
PG13	0.000243	0.000224	v_{PGK}	2.24	2.24
PG3	0.0773	0.0898	v_{PGLM}	2.24	2.24
PG2	0.0113	0.0131	v_{ENO}	2.24	2.24
PEP	0.017	0.0201	v_{PK}	2.24	2.24
PYR	0.060301	0.0603	v_{LDH}	2.016	2.016
LAC	1.36	1.36	v_{AMP}	0.014	0.015
NAD	0.0589	0.0589	v_{APK}	0.0	0.0274
NADH	0.0301	0.0301	v_{PYR}	0.224	0.224
AMP	0.0867	0.0930	v_{LAC}	2.016	2.016
ADP	0.29	0.245	$\mathbf{v_{ATP}}$	**3.36**	**2.24**
ATP	1.6	1.067	v_{NADH}	0.224	0.224
Phos	2.5	3.619	v_{GLUin}	1.12	1.12
H$^+$	0.000103	0.000103	v_{AMPin}	0.014	0.014
H$_2$O	1.0	1.0	v_{H^+}	2.688	2.688
			v_{H_2O}	0.0	0.0

from Eq. (10.4). If the reaction is close to equilibrium, then the denominator can be close to zero. If this is the case, the PERC is effectively indeterminable, but we know it is fast. The PERC can then be fixed to a large value and, in almost all cases, will correspond to dynamics that are too fast to be of interest.

Node maps   These maps show all the flows of mass in and out of a node (a compound) in the network. Some of the key nodes in the glycolytic system are shown in Figure 10.3.

The rate constants of the links into and out of a node can differ in their magnitude. They give us a measure of the spectrum of response times associated with a node. It is hard to excite some of the rapid dynamics associated with a node by perturbing a boundary flux; thus, such rapid

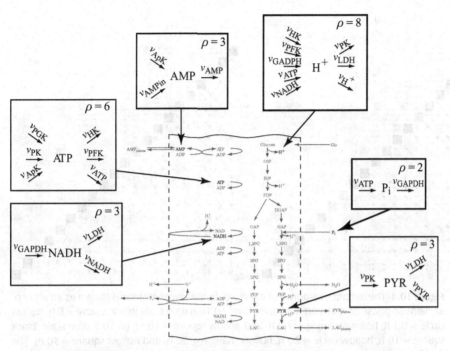

Figure 10.3 Node maps for glycolytic intermediates. The node maps for AMP, NADH, ATP, P$_i$, PYR, and H$^+$. The flows in and out balance in the steady state.

motions are not often observed in dynamical simulations around a steady state.

10.4 Simulating mass balances: biochemistry

We have now specified the system, its contents, its steady state, and all numerical values needed to simulate dynamic responses. We can now introduce **S** into the dynamic mass balances (Eq. (1.1)) and simulate the response to perturbations. Such simulations can be done for perturbations in the energy or redox load, in environmental parameters like the external PYR and LAC concentrations, or in the influx of glucose or AMP. A number of cases are outlined in the exercises at the end of the chapter and can be implemented with the Mathematica workbook provided.

We now simulate the dynamic mass balances that describe the time profiles of concentrations. We will be studying individual concentrations and the reactions that affect their levels; therefore, we are looking at the system at the biochemical level.

Response to an increased k$_{ATP}$ at a constant glucose input rate   Continuing the discussion from Sections 8.3 and 8.4, we focus here on a

(a)   (b)

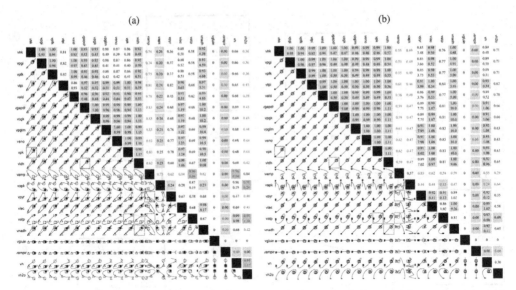

Figure 10.4 The dynamic response of the fluxes in the glycolytic model shown as an array of all pairwise phase portraits on two time scales. (a) 10 h time scale (black square = 0 h, hollow circle = 0.1 h, hollow triangle = 1.0 h, and hollow square = 10 h). (b) 50 h time scale (black square = 10 h, hollow circle = 20 h, hollow triangle = 30 h, and hollow square = 50 h). The numbers above the diagonal are the correlation coefficient and slope for the corresponding phase portrait below the diagonal. The boxed phase portraits are detailed in Figure 10.5.

perturbation in the ATP load, where we increase the k_{ATP} parameter by 50% at $t = 0$ and simulate the dynamic response to a new steady state. This perturbation reflects a change in the rate of usage of ATP. Based on the distribution of rate constants in Table 10.5, we estimate that we have roughly three time scales of interest: <1 h, ~10 h, and ~50 h.

Systems versus component variables   The fluxes are the systems variables and we look at them first. Since the input rate of glucose is the same throughout, the fluxes end up at their initial values but the concentrations will not. This statement is important; the concentrations have to adjust to the requirements of the system to balance its fluxes in a steady state. Thus, in this sense, the concentrations are subordinate variables to the fluxes. This notion is important in systems biology. We are dealing with *systems* properties, not *component* properties as one would in carrying out a single reaction *in vitro*.

The dynamic response of the fluxes   In this simulation, the rate constant for ATP usage was changed. The start and end point values for the fluxes are given in Table 10.6. To get a full dynamic view of the response, we prepare tiled phase portraits of the fluxes on the 10 and 50 h time scale (Figure 10.4). These tiled arrays of phase portraits are informative and

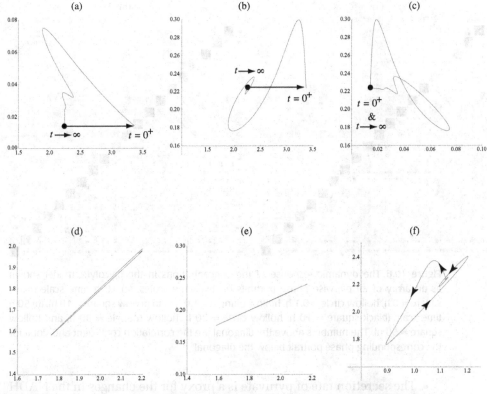

Figure 10.5 The dynamic response of key fluxes. Detailed pairwise phase portraits: (a) v_{ATP} versus v_{AMP}. (b) v_{ATP} versus v_{PYR}. (c) v_{AMP} versus v_{PYR}. (d) v_{GAPDH} versus v_{LDH}. (e) v_{LDH} versus v_{NADH}. (f) v_{HK} versus v_{PK}. The fluxes are in units of mM/h. The perturbation is reflected in the instantaneous move of the flux state from the initial steady state to an unsteady state, as indicated by the arrow placing the initial point at $t = 0^+$. The system then returns to its steady state at $t \rightarrow \infty$.

help obtain a comprehensive view of the response of the glycolytic fluxes to increased rates of energy use. Some observations are as follows:

- The dynamic changes in the fluxes show a high degree of coordination. Thus, the effective dynamic order of the response, on these time scales is much lower than the dynamic dimension of the full model. The response is effectively on two of the slowest time scales.
- Three fluxes determine most of the dynamic behavior on these slowest time scales. They are v_{ATP}, v_{AMP}, and v_{PYR}. Note that these three fluxes are associated with the three pathway vectors that span the null space. This issue is discussed in more detail in the next section. The detailed dynamic phase portraits for these three fluxes are shown in Figure 10.5a–c.
- The response shows a damped oscillation. This feature is not a result of any special regulatory mechanism, as the model being simulated is solely based on stoichiometry and elementary rate laws.

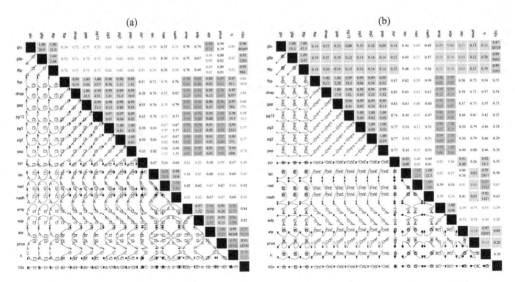

Figure 10.6 The dynamic response of the concentrations in the glycolytic model shown as an array of all pairwise phase portraits on two time scales: (a) 10 h time scale (black square = 0 h, hollow circle = 0.1 h, hollow triangle = 1.0 h, and hollow square = 10 h). (b) 50 h time scale (black square = 10 h, hollow circle = 20 h, hollow triangle = 30 h, and hollow square = 50 h). The numbers above the diagonal are the correlation coefficient and slope for the corresponding phase portrait below the diagonal.

- The secretion rate of pyruvate is a proxy for the changes in the NADH load, as discussed earlier. The perturbations in the pyruvate secretion rate thus represent how the change in the rate of energy usage influences the redox potential production by glycolysis.
- The perturbation leads to AMP secretion rates that are above its input rate. Since the ATP levels drop initially in response to the increased rate of usage, the level of AMP increases, and it exits the system. This AMP secretion is the long-term response of the system.
- The response of the dehydrogenases is rapid and balanced, as shown in Figures 10.5d and e. There is no change in the redox load, so these fluxes have to balance while the system meets the increased use rate of ATP.
- The initial reaction is reduced flux through HK, while PK increases; see Figure 10.5f. ATP drops initially and leads to this short-term response. Afterward, both fluxes drop and then increase to reach the original steady state.

Concentrations   The tiled dynamic phase portraits for the concentrations are shown in Figure 10.6. The dynamics of the concentrations are "richer" than those of the fluxes. The reason is that the changes in the concentrations to a flux perturbation are greater than those needed in the fluxes to reach the new steady state. The eventual state for the fluxes will be the

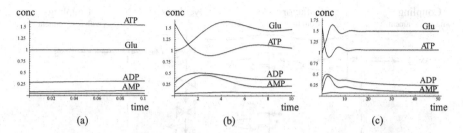

Figure 10.7 The dynamic response of the glycolytic model on three key time scales: (a) less than an hour; (b) 10 h; (c) 50 h. The parameter k_{ATP} is increased by 50% at $t = 0$.

same, but the concentrations will reach different levels. Some observations of these phase portraits are as follows:

- Glucose, G6P, and F6P move as a pool (note that the slope in the phase portrait of G6P and F6P is $K_{eq} = 0.41 = 1/2.44$). This pool is "soft," as its size changes after the initial equilibration step.
- In contrast, the time-invariant pools are "hard" as their size never changes. The time-invariant of $p_{11} = NADH + NAD$ is reflected in the motion along a $-45°$ line with a perfect correlation coefficient.

The response is essentially on two time scales, as illustrated in Figure 10.7.

- Similarly, the trioses and the phosphoglycerates move in a highly coordinated fashion, as do ADP and AMP, PYR, LAC, P_i, and NADH.
- The perturbation leads to AMP secretion rates that are above its input rate, leading to a slow decay in the AMP + ADP + ATP inventory. This long-term response has a determining effect on the property ratios discussed below.

10.5 Pooling: towards systems biology

We now take a look at the biochemical features of this pathway to formulate meaningful pooling of the concentrations to form aggregate variables. The formulation of these quantities allows for a systems interpretation of the dynamic responses rather than the chemical interpretation that is achieved by looking at individual concentrations and fluxes. The analysis is organized around the three cofactor coupling features of glycolysis. A symbolic version of the compounds, shown in Figure 10.8, can help us in this process.

High-energy phosphate bond trafficking (Figure 10.9)   There are four kinases (HK, PFK, PGK, and PK) in the pathway. With respect to the

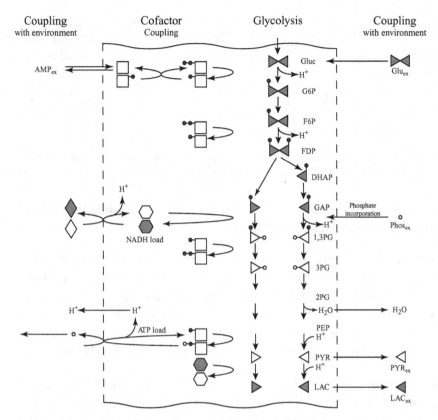

Figure 10.8 A schematic of glycolysis that symbolically shows how redox (shaded = reduced, clear = oxidized) energy (see Figure 8.4 for symbology) and inorganic phosphate (open circle) incorporation are coupled to the pathway. The closed circles represent phosphates that cycle between the adenylates and the glycolytic intermediates. A triangle is used to symbolize a three-carbon compound.

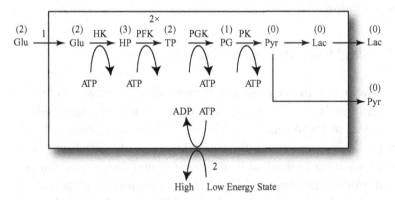

Figure 10.9 High-energy bond trafficking in glycolysis. The numbers above the arrows are the fluxes. The 2× indicates doubling in flux where a hexose is split into two trioses. The numbers in parentheses above the compounds represent their high-energy bond value.

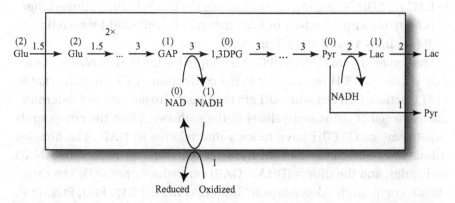

Figure 10.10 Redox trafficking in glycolysis. The numbers above the arrows are the fluxes in a steady state. The 2× indicates doubling in flux where a hexose is split into two trioses. If 1.5 flux units enter glycolysis and the redox load is one, then two flux units of lactate and one of pyruvate are formed. The numbers in parentheses above the compounds represent their redox value.

potential to generate high-energy phosphate bonds, PYR and LAC are in a "ground state," incapable of generating such bonds. The phosphoglycerates (PG3 + PG2 + PEP) are capable of generating one high-energy bond, the triose phosphates (2FDP + DHAP + GAP + DPG13) are capable of generating two, the hexose phosphates (G6P + F6P) three, and glucose two. Thus, the high-energy inventory (occupancy) in glycolysis is

$$p_1 = 2\text{Gluc} + 3(\text{G6P} + \text{F6P})$$
$$+ 2(2\text{FDP} + \text{DHAP} + \text{GAP} + \text{DPG13})$$
$$+ (\text{PG3} + \text{PG2} + \text{PEP}).$$

The intermediates that have no high-energy phosphate bond value are

$$p_2 = \text{PYR} + \text{LAC}. \tag{10.5}$$

The high-energy bonds made in glycolysis are then stored in the form of ATP, which in turn is used to meet energy demands. As we saw in Chapter 2, the adenylate phosphates can be described by the two pools

$$p_3 = \text{ADP} + 2\text{ATP} \quad \text{and} \quad p_4 = 2\text{AMP} + \text{ADP} \tag{10.6}$$

that represent the *capacity* and *vacancy* of high-energy bonds on the adenylates. The addition of the two gives us the *capacity*, as we saw in Chapter 2.

Redox trafficking (Figure 10.10) The net flow of glucose to two lactate molecules is neutral with respect to generation of redox potential. If there is a separate redox load (see line 17 in Table 10.5) then some of the NADH that is produced by GAPDH is used for other purposes than to reduce PYR to LAC via LDH. Then, in order to balance the fluxes in the pathway, there will be reduced production of LAC that will be reflected by secretion of PYR (see lines 14 and 15 in Table 10.5).

The conversion of GAP to DPG13 is coupled to the NADH/NAD cofactor pair via the GAPDH reaction, as is the conversion of PYR to LAC via the LDH reaction. GAPDH and LDH are the two hydrogenases that determine the exchange of redox equivalents in the pathway. Thus, the compounds "upstream" of GAPDH have redox value relative to NAD. The hexoses (Gluc, G6P, F6P, and FDP) all have the potential to reduce two NAD molecules, and the trioses (DHAP, GAP) can reduce one NAD. The three-carbon compounds "downstream" from GAPDH (DPG13, PG3, PG2, PEP, and PYR) have no redox value. The conversion of PYR to LAC using NADH gives LAC a redox value of one NADH. The two dehydrogenases are reversible enzymes. Note that this assignment of redox value is dependent on the environment in which the pathway operates, Figure 10.1, as defined by the systems boundary drawn and the inputs and outputs given.

The total redox inventory (occupancy) of the glycolytic intermediates is thus given by

$$p_5 = 2(\text{Gluc} + \text{G6P} + \text{F6P} + \text{FDP}) + (\text{DHAP} + \text{GAP}) + \text{LAC}.$$

By the same token, the intermediates in an oxidized state are

$$p_6 = \text{DPG13} + \text{PG3} + \text{PG2} + \text{PEP} + \text{PYR}.$$

The carrier of the redox potential is NADH; we thus define

$$p_7 = \text{NADH} = \text{N}^+$$

as the occupancy of redox potential. Clearly, then, NAD would be the vacancy state, denoted by N^-. However, in this case, $p_{11} = \text{N}^+ + \text{N}^-$ will be the total capacity to carry redox potential by this carrier. It is a time-invariant pool in this model, since the cofactor moiety cannot enter or leave the system.

The trafficking of phosphate groups (Figure 10.11)   Glycolysis generates a net of two ATP per glucose consumed; two ATP molecules are spent and four are generated. In the process, GAPDH incorporates the net of two inorganic phosphate groups that are needed to generate the net two ATP from ADP. These features lead to distinguishing two types of phosphate group in glycolysis: those that were recycled between the glycolytic inter-mediates and the adenylate carrier or being used through the ATP load reaction

$$p_8 = \text{G6P} + \text{F6P} + 2\text{FDP} + \text{DHAP} + \text{GAP} + \text{DPG13} + \text{ADP} + 2\text{ATP}$$

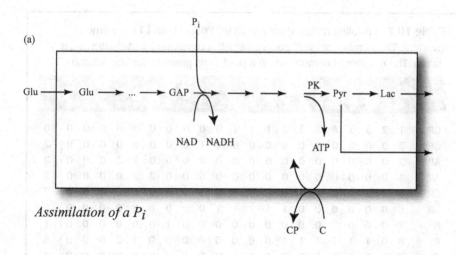

Assimilation of a Pᵢ

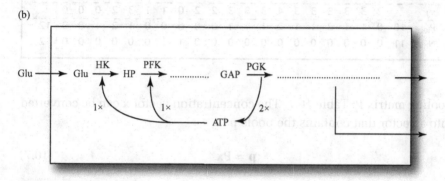

Internal cycling or
"external use" of ~Pᵢ

Figure 10.11 The trafficking of phosphate groups in glycolysis. (a) The incorporation of inorganic phosphate that then gets passed onto the molecule C, where it creates a high-energy phosphate bond (in the example simulated in this chapter, the load is simply a hydrolysis reaction and there is no C molecule). (b) The cycling of high-energy phosphate groups between the adenylate carrier and the glycolytic intermediates. The relative steady-state flux values are indicated by $1\times$ and $2\times$.

and those that were incorporated

$$p_9 = \text{DPG13} + \text{PG3} + \text{PG2} + \text{PEP}.$$

If the inorganic phosphate is added to the sum of these two pools, then we get the total phosphate inventory, which is the second time-invariant, p_{10}.

The pooling matrix The inspection of the biochemical properties of the glycolytic pathway leads to the definition of a series of pools. This biochemical insight can be formally represented mathematically using a

Table 10.7 A pooling matrix **P** for glycolysis. Pools 10 and 11 are time invariant. The number of "soft" pools in which a compound participates, π_j, is shown. The number of compounds in a pool, ρ_i, is given in the last column

	#	Glu	G6P	F6P	FBP	DHAP	GAP	PG13	PG3	PG2	PEP	PYR	LAC	NAD	NADH	AMP	ADP	ATP	Pi	H+	H2O	ρ_i
GP+	1	2	3	3	4	2	2	2	1	1	1	0	0	0	0	0	0	0	0	0	0	10
GP−	2	0	0	0	0	0	0	0	0	0	0	1	1	0	0	0	0	0	0	0	0	2
AP+	3	0	0	0	0	0	0	0	0	0	0	0	0	0	0	1	2	0	0	0	0	2
AP−	4	0	0	0	0	0	0	0	0	0	0	0	0	0	0	0	2	1	0	0	0	2
GR+	5	2	2	2	2	1	1	0	0	0	0	0	1	0	0	0	0	0	0	0	0	7
GR−	6	0	0	0	0	0	0	1	1	1	1	1	0	0	0	0	0	0	0	0	0	5
N+	7	0	0	0	0	0	0	0	0	0	0	0	0	0	1	0	0	0	0	0	0	1
P+	8	0	1	1	2	1	1	1	0	0	0	0	0	0	0	0	1	2	0	0	0	8
P−	9	0	0	0	0	0	0	1	1	1	1	0	0	0	0	0	0	0	0	0	0	4
π_j		2	3	3	3	3	3	4	3	3	3	2	2	0	1	1	3	2	0	0	0	
P$_{tot}$	10	0	1	1	2	1	1	2	1	1	1	0	0	0	0	1	2	1	0	0	0	12
N$_{tot}$	11	0	0	0	0	0	0	0	0	0	0	0	0	1	1	0	0	0	0	0	0	2

pooling matrix **P**; Table 10.7. The concentration vector **x** can be converted into a vector that contains the pools **p** by

$$p = Px \qquad (10.7)$$

as a post-processing step (see Equation (3.4)).

The table sums up the nonzero elements in a row (ρ_i) and in a column (π_j). The number of compounds that make up a pool is given by ρ_i, while π_j gives the number of pools in which a compound participates. Thus, π_j tells us how many aggregate metabolic properties in which a compound participates. For instance, G6P has a glycolytic energy value, glycolytic redox value, and has a high-energy phosphate group. G6P is thus in three ($\pi_{G6P} = 3$) different glycolytic pools.

This multifunctionality of a compound is common in biochemical reaction networks. This feature makes it hard to untangle the individual functions of a compound (recall the "tangle of cycles" notion from Chapter 8). This coupling of individual functions of a molecule (or a node) in a network is a foundational feature of systems biology.

The reactions that move the pools The dynamic mass balances can be multiplied by the matrix **P** to obtain dynamic balances on the pools, as

$$\frac{dp}{dt} = P\frac{dx}{dt} = PSv(x). \qquad (10.8)$$

The matrix **PS** (see Table 10.8) therefore tells us which fluxes move each of the pools.

Table 10.8 also contains the summation of the nonzero elements in a row (ρ_i) and a column (π_j). The former gives the number of fluxes that fill or drain a pool. Thus, when drawing the system in terms of pools (see Figure 10.12), p_i is the number of links in a pool. Notice that the last two rows have no nonzero entries, since these are time-invariant pools.

The latter number, π_j, is the number of pools that a flux moves. Many of the π_j are zero, and thus these reactions do not appear in Figure 10.12. Conversely, some fluxes drain and fill many pools. PK moves six pools, while GAPDH moves four pools.

Graphical representation   The glycolytic system can now be laid out in terms of the pools (see Figure 10.12). This layout shows how the fluxes interconnect the pools and how the loads pull on these pools and lead to the movement of mass among the pools. This diagram is akin to a process flow chart. The pools sizes, the steady-state flux through them, and their response times, and the ratio of the sizes to the net fluxes, are shown in Table 10.8. Note that we have now moved our point of view from the individual molecules to a view where aggregate redox, energy, and phosphate values are displayed.

Dynamic behavior of the pools   The pooling matrix **P**, defined in Table 10.7, can be used to compute the pools by $p(t) = \mathbf{P}x(t)$. The time-dependent responses of the pools can then be studied. The dynamic responses to a sudden increase in ATP load of the phosphate-containing pools are shown in Figure 10.13. These pools go through well-defined changes:

- The sudden change in the ATP usage rate creates a flux imbalance that leads to drainage of the AP^+ and P^+ pools that is mirrored in the build-up of the AP^- pool.
- Counterintuitively, taken together, the activities of the kinases lead to an increase in the GP^+ and GR^+ pools.
- There is a slow drainage of the AP^- pool that corresponds to the exit of AMP from the system, which is reflected in an increase in the P^- and the GR^- pools.
- There is a dynamic interaction with the NADH pool leading to some drainage of redox potential.

Table 10.8 The fluxes that move the pools. This table is obtained from the product **PS**. The time-invariant pools are in the last two rows. The number of reactions ρ_l that move a pool is shown, as are the pool size, the steady-state flux in and out of the pool, and its turnover time. The number of pools that a reaction moves, π_J, is also given

	V_{NADH}	V_{PGI}	V_{PFK}	V_{TPI}	V_{ALD}	V_{GAPDH}	V_{PGK}	V_{PGLM}	V_{ENO}	V_{PK}	V_{LDH}	V_{AMP}	V_{APK}	V_{PYR}	V_{LAC}	V_{ATP}	V_{NADH}	V_{Gluin}	V_{AMPin}	V_{H+}	V_{H_2O}	ρ_l	Size (mM)	Net stst flux (mM/h)	τ (h)
GP+	1	0	1	0	0	0	0	0	0	−1	0	0	0	0	0	0	0	2	0	0	0	5	2.70	4.48	0.60
GP−	0	0	0	0	0	0	−1	0	0	1	0	0	0	−1	−1	0	0	0	0	0	0	3	1.42	2.24	0.63
AP+	−1	0	−1	0	0	0	1	0	0	1	0	0	0	0	0	−1	0	0	0	0	0	5	3.49	4.48	0.78
AP−	1	0	1	0	0	0	−1	0	0	−1	0	−1	0	0	0	1	0	0	1	0	0	5	0.46	4.51	0.10
GR+	0	0	0	0	0	−1	0	0	0	0	1	0	0	0	−1	0	0	2	0	0	0	4	3.69	4.26	0.87
GR−	0	0	0	0	0	1	0	0	0	0	−1	0	0	−1	0	0	0	0	0	0	0	3	0.17	2.24	0.07
N+	0	0	0	0	0	1	0	0	0	0	−1	0	0	0	0	0	−1	0	0	0	0	3	0.03	2.24	0.013
P+	0	0	0	0	0	0	0	0	0	1	0	0	0	0	0	−1	0	0	0	0	0	2	3.76	2.24	1.68
P−	0	0	0	0	0	1	0	0	0	−1	0	0	0	0	0	0	0	0	0	0	0	2	0.11	2.24	0.05
π_J	3	0	3	0	0	4	3	0	0	6	3	1	0	2	2	3	1	2	1	0	0	0	0.11	2.24	0.05
P$_{tot}$	0	0	0	0	0	0	0	0	0	0	0	0	0	0	0	0	0	0	0	0	0	0	6.36		∞
N$_{tot}$	0	0	0	0	0	0	0	0	0	0	0	0	0	0	0	0	0	0	0	0	0	0	0.09		∞

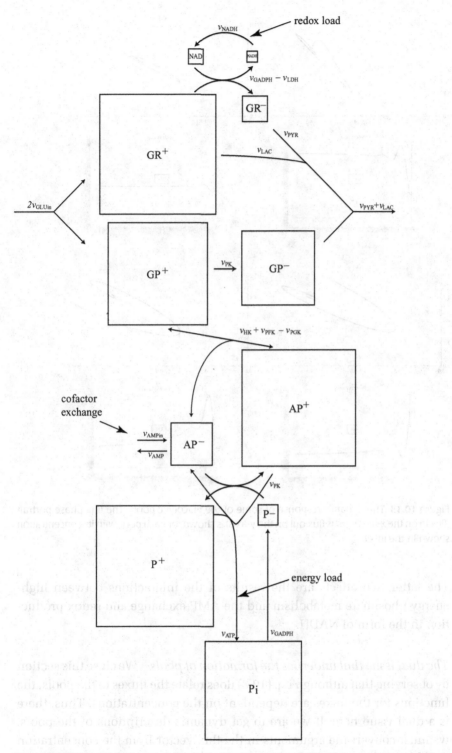

Figure 10.12 A pool-flux map of glycolysis. The pools from Table 10.7 are shown as well as the fluxes that fill and drain them (Table 10.8). The relative sizes of the boxes indicate the relative concentration of a pool.

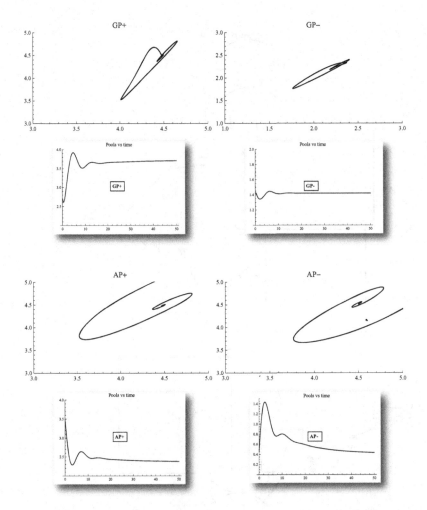

Figure 10.13 The dynamic response of some of the glycolytic pools. The flux phase portrait (flux in on the x-axis versus flux out on the y-axis) is shown for each pool, with its concentration shown in the inset.

The latter two effects are the results of the interactions between high-energy phosphate metabolism and the AMP exchange and redox production in the form of NADH.

The dual issue that underlies the formation of pools   We close this section by observing that although Eq. (10.8) does relate the fluxes to the pools, the functions for the fluxes are dependent on the concentrations. Thus, there is a dual issue here. If we are to get dynamic descriptions of the pools, we must convert the arguments in the flux vector from the concentration variables to the pools themselves. This problem is mathematically difficult to analyze and is addressed elsewhere [90].

10.6 Ratios: towards physiology

We now take a look at the conjugate pools (i.e., the corresponding high and low states) of various metabolic properties of the glycolytic system to form property ratios. The energy charge is one such ratio and was introduced in Chapter 2. The formulation of these quantities allows a physiological interpretation of the dynamic responses.

The notion of a charge is related to the relative size of the conjugate pools. A ratio is formed as

$$\text{property ratio} = \frac{\text{high}}{\text{high} + \text{low}} = \frac{\text{high}}{\text{total}}. \tag{10.9}$$

The ratios r_i are computed from the conjugates p_i. We can then graph the ratios to interpret transient responses from a metabolic physiological point of view. The total pool size may change on a different time scale than the interconversion between the high and low forms. Any time-scale hierarchy in such responses is of interest, since most physiological responses involve multiscale analysis.

Energy charges We can define an energy charge for glycolysis as the ratio

$$r_1 = \frac{p_1}{p_1 + p_2}, \tag{10.10}$$

which is an analogous quantity to the adenylate energy charge:

$$r_2 = \frac{p_3}{p_3 + p_4} = \frac{2\text{ATP} + \text{ADP}}{2(\text{ATP} + \text{ADP} + \text{AMP})}. \tag{10.11}$$

Thus, we have two ratios that describe metabolic physiology in terms of the energy charge in the glycolytic intermediates and on the adenylate phosphates.

These charge parameters (r_1 and r_2) can vary between zero and unity. If the ratio is close to unity, then the energy charge is high, and vice versa. The glycolytic pathway thus transfers metabolic energy equivalents from glucose to ADP. The energy charge of the glycolytic intermediates and NAD are quantities that describe how this process takes place.

Redox charges We can define the redox charge in glycolysis as

$$r_3 = \frac{p_5}{p_5 + p_6}. \tag{10.12}$$

We note that three times the denominator in Eq. (10.12) is the total carbon inventory in glycolysis. In an analogous fashion, we can define the redox

Table 10.9 The numerical values of key ratios in glycolysis before and after the increased rate of ATP use

Ratio	#	$t=0$	$t=50$
Glycolytic energy charge	1	0.656	0.723
Adenylate energy charge	2	0.883	0.847
Glycolytic redox charge	3	0.957	0.962
NADH redox charge	4	0.338	0.338
Phosphate recycle ratio	5	0.972	0.955

state on the NAD carrier as

$$r_4 = \frac{\text{NADH}}{\text{NADH} + \text{NAD}} = \frac{p_7}{p_{10}}. \qquad (10.13)$$

These ratios will have identical interpretations as the energy charges. Glycolysis will move redox equivalents in glucose onto NADH that then become the conduit of redox equivalent to other processes in a cell, here represented by the NADH load function.

The state of the phosphate groups   We can define the state of the phosphates as

$$r_5 = \frac{p_9}{p_8 + p_9}$$

to get the fraction of phosphate that has been incorporated and is available for recycling or to meet the ATP demand function.

Operating diagrams   The combination of the conjugate pools can be used to develop the pool diagram of Figure 10.12 further and form effectively an operating diagram for glycolysis; Figure 10.14. This diagram represents the glycolytic system as a set of interconnected charges of various properties and is a physiological point of view.

Dynamic responses of the ratios   The ratios can easily be computed from the time profiles of the pools (Table 10.9). The response of the ratios is shown on the glycolytic operating diagram in Figure 10.14. We now see the simplicity in the dynamic reaction of the glycolytic system to the increased rate of ATP usage.

- The energy charge experiences a significant drop initially, but then recovers over a long time period, due to the reduction in the total adenosine phosphate ATP + ADP + AMP pool.

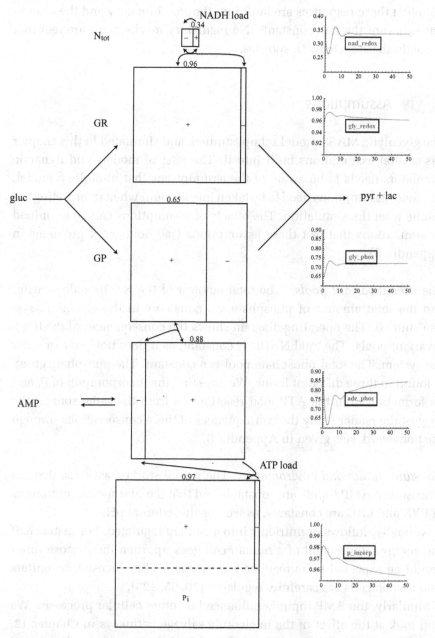

Figure 10.14 An operating diagram for the glycolytic systems and the dynamic response of the glycolytic property ratios.

- The phosphate charge has a similar but much less pronounced adjustment.
- The energy charge of glycolysis increases.
- There are minor ripple effects through the redox pools.

Note that these responses are built into the stoichiometry and the numerical values of the rate constants. No regulatory mechanisms are required to obtain these systemic responses.

10.7 Assumptions

The glycolytic MASS model set up, studied, and simulated in this chapter has several assumptions built into it. The user of models and dynamic simulators needs to be aware of the assumptions that underlie a model, and these assumptions need to be taken into account when interpreting the results from the simulation. The effects of assumptions can be examined by simulations that test these assumptions (see homework problems in Appendix B).

The time-invariant pools The total amount of the NADH redox carrier and the total amount of phosphate are constants in the glycolytic system studied. The operating diagram shows the consequences of the time-invariant pools. The total NADH is constant, as it does not leave or enter the system. The total phosphate pool is a constant. The phosphate group is found in three different forms. We note that the incorporation of P_i and its formation from the ATP load reaction are from the same source. We suggest the reader study the consequences of these conservations through the homework sets given in Appendix B.

Constant inputs and environment The model studied assumes that the glucose and AMP inputs are constants and that the plasma concentrations of PYR and LAC are constants, as well as the external pH.

Normally, inflows of nutrients into a cell are regulated. For instance, if the energy requirement of a human cell goes up, then the glucose input would be expected to be regulated to increase. The glucose transporters into human cells are carefully regulated [70, 95, 123].

Similarly, the AMP input is influenced by other cellular processes. We will look at the effect of the nucleotide salvage pathways in Chapter 12. Through the simulations performed in this chapter, we have discovered that this input rate is important for the long-term function of the system and for the determination of its energy state. The salvage pathways have an important physiological role, and many serious inborn errors of metabolism are associated with genetic defects in this nucleotide salvage process [28, 76].

The composition of the plasma is variable. For instance, the normal range of pyruvate concentration in plasma is 0.06 to 0.11 mM, and that for lactate is 0.4 to 1.8 mM. Here, we did fix these conditions to one condition. One can simulate the response to changes in plasma concentrations (see homework 10.6 and 10.7).

Constant volume and charge neutrality As discussed in Chapter 1 (Table 1.2), there are several fundamental assumptions that underlie the formulation of the dynamic mass balances. These include charge neutrality of the compounds and electroneutrality of the interior and exterior of the cell. The simulator studied here assumed neutral molecules and no osmotic effects on the volume of the system.

Regulation There were no regulatory mechanisms incorporated into this simulator. In Chapter 14 we will discuss how regulated enzymes are described in MASS models.

10.8 Summary

➤ Initial draft MASS models can be obtained from using measured concentration values, elementary reactions, and associated mass action kinetics. These first draft kinetic models can be used as a scaffold to build more complicated models that include regulatory effects and interactions with other pathways.

➤ Dynamic simulations can be performed for perturbations in environmental parameters, and the responses can be examined in terms of the concentrations and the fluxes. Tiled phase portraits are useful to get an overall view of the dynamic response.

➤ A metabolic map can be analyzed for its stoichiometric texture to assess consequences of cofactor coupling. Such reduction of the biochemical network helps define pools that are physiologically meaningful from a metabolic perspective and are context dependent.

➤ Some of the responses, namely damped oscillatory behavior, are built into the topological features of a network and require no regulatory action.

➤ The raw output of the simulation can be post-processed with a pooling matrix that allows the pools and their ratios to be graphed to obtain a deeper interpretation of dynamic responses.

➤ MASS models, post-processing, and analysis of responses allows study at three levels: (1) biochemistry, (2) systems biology, and (3) physiology.

## CHAPTER 11

Coupling pathways

In Chapter 10 we formulated a MASS model of glycolysis. We took a linear pathway and converted it into an open system with defined inputs and outputs, formed the dynamic mass balances, and then simulated its response to increased rate of energy use. In this chapter, we will show how one can build a dynamic simulation model for two coupled pathways that is based on an integrated stoichiometric scaffold for the two pathways. We start with the pentose pathway and then couple it to the glycolytic model from Chapter 10 to form a simulation model of two pathways to study their simultaneous dynamic responses.

11.1 The pentose pathway

The pentose pathway originates from G6P in glycolysis (Figure 11.1). The pathway is typically thought of as being comprised of two parts: the oxidative and the nonoxidative branches.

The oxidative branch G6P undergoes two oxidation steps, including decarboxylation, releasing CO_2, leading to the formation of one pentose and two NADPH molecules. These reactions are called the *oxidative branch* of the pentose pathway. The branch forms two NADPH molecules that are used to form glutathione (GSH) from an oxidized dimeric state, GSSG, by breaking a disulfite bond. GSH and GSSG are present in high concentrations, and thus buffer the NADPH redox charge (recall the discussion of the creatine phosphate buffer in Chapter 8). The pentose formed, R5P, can be used for biosynthesis. We will discuss the connection of R5P with the salvage pathways in Chapter 12.

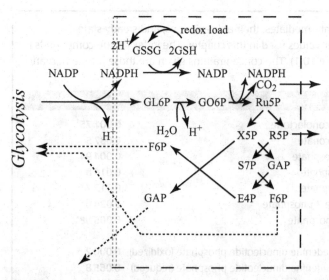

Figure 11.1 The pentose pathway. The reaction schema, cofactor interactions, and environmental exchanges.

The nonoxidative branch If the pentose formed by the oxidative branch is not used for biosynthetic purposes, it undergoes a number of isomerization, epimerization, transaldolation, and transketolation reactions that lead to the formation of F6P and GAP. Specifically, two F6P and one GAP that return to glycolysis come from three pentose molecules (i.e., $3 \times 5 = 15$ carbon atoms go to $2 \times 6 + 3 = 15$ carbon atoms). This part of the pathway is the *nonoxidative branch*, and it is comprised of a series of reversible reactions, while the oxidative branch is irreversible.

The overall reaction schema When no pentose is used by other pathways, the overall flow of carbon in the pentose pathway can be described by

$$3\text{G6P} \rightarrow 2\text{F6P} + \text{GAP} + 3\text{CO}_2. \tag{11.1}$$

The input and output from the pathway are glycolytic intermediates. Two redox equivalents of NADPH (or GSH) are produced for every G6P that enters the pathway.

Metabolic functions The function of the pentose pathway is to produce redox potential in the form of NADPH and a pentose phosphate that is used for nucleotide synthesis. Also, a four-carbon intermediate, E4P, is used as a biosynthetic precursor for amino acid synthesis. Exchange reactions for R5P and E4P can then be added to complete the system, as well as a redox load on NADPH.

Table 11.1 Pentose pathway intermediates, their abbreviations, and steady-state concentrations for the parameter values used in this chapter. The index on the compounds is added to that for glycolysis (Table 10.1). The concentrations given are those for the human red blood cell

#	Abbreviation	Intermediates/cofactors	Concentration (mM)
21	GL6P	6-Phosphogluconolactone	0.001 75
22	GO6P	6-Phosphoglyconate	0.037 5
23	Ru5P	Ribulose 5-phosphate	0.004 94
24	X5P	Xylose 5-phosphate	0.014 8
25	R5P	Ribose 5-phosphate	0.012 7
26	S7P	Sedoheptulose 7-phosphate	0.024 0
27	E4P	Erythrose 4-phosphate	0.005 08
28	NADP	Nicotinamide adenine dinucleotide phosphate (oxidized)	0.000 2
29	NADPH	Nicotinamide adenine dinucleotide phosphate (reduced)	0.065 8
30	GSH	Glutathione (reduced)	3.2
31	GSSG	Glutathione (oxidized)	0.12
32	CO_2	Carbon dioxide	1.0 (arbitrary)

Table 11.2 Elemental composition of the intermediates of the pentose pathway. The total pools of glutathione (G_{tot}) and NADPH (NP_{tot}) are indicated

	G6P	F6P	GAP	GL6P	GO6P	RU5P	X5P	R5P	S7P	E4P	NADP	NADPH	GSH	GSSG	CO_2	H	H_2O
C	6	6	3	6	6	5	5	5	7	4	0	0	10	20	1	0	0
H	11	11	5	9	10	9	9	9	13	7	0	1	17	32	0	1	2
O	9	9	6	9	10	8	8	8	10	7	0	0	6	12	2	0	1
P	1	1	1	1	1	1	1	1	1	1	0	0	0	0	0	0	0
N	0	0	0	0	0	0	0	0	0	0	0	0	3	6	0	0	0
S	0	0	0	0	0	0	0	0	0	0	0	0	1	2	0	0	0
NADP	0	0	0	0	0	0	0	0	0	0	1	1	0	0	0	0	0
G_{tot}	0	0	0	0	0	0	0	0	0	0	0	0	1	2	0	0	0
NP_{tot}	0	0	0	0	0	0	0	0	0	0	1	1	0	0	0	0	0

The biochemical reactions The pentose pathway has 12 compounds (Table 11.1). Their elemental composition is given in Table 11.2. The pathway has 11 reactions (Table 11.3). We will consider the removal of R5P, but not of E4P, in this chapter. An exchange reaction for E4P can be added if desired.

Table 11.3 Pentose pathway enzymes and transporters, their abbreviations, and chemical reactions. The index on the reactions is added to that for glycolysis (Table 10.5). The reactions of the oxidative branch are irreversible, while those of the nonoxidative branch are reversible

	Abbreviation	Enzymes/transporter/load	Elementally balanced reaction
22	G6PDH	G6P dehydrogenase	$G6P + NADP \rightarrow GL6P + NADPH + H$
23	PGLase	6-Phosphogluconolactonase	$GL6P + H_2O \rightarrow GO6P + H$
24	GL6PDH	GO6P dehydrogenase	$GO6P + NADP \rightarrow RU5P + NADPH + CO_2$
25	R5PE	X5P epimerase	$RU5P \rightleftharpoons X5P$
26	R5PI	R5P isomerase	$RU5P \rightleftharpoons R5P$
27	TKI	Transketolase I	$X5P + R5P \rightleftharpoons S7P + GAP$
28	TKII	Transketolase II	$X5P + E4P \rightleftharpoons F6P + GAP$
29	TALA	Transaldolase	$S7P + GAP \rightleftharpoons E4P + F6P$
30	GSSGR	Glutathione reductase	$GSSG + NADPH + H \rightleftharpoons 2GSH + NADP$
31	GSHR	Glutathione oxidase	$2GSH \rightleftharpoons GSSG + 2H$
32	CO_2 exch	Freely exchanging CO_2	$CO_{2,in} \rightarrow CO_{2,out}$

Stoichiometric matrix for the pentose pathway The properties of the stoichiometric matrix are shown in Table 11.4. All the reactions are elementally balanced except for the exchange reactions. The matrix has dimensions of 17×17 and a rank of 15. It thus has a two-dimensional null space and a two-dimensional left null space.

The two pathway vectors that span the null space are shown towards the bottom of Table 11.4. The first pathway, $\mathbf{p}_1$, corresponds to the schema shown in Eq (11.1), representing the use of the pentose pathway out of and back into glycolysis. The second pathway, $\mathbf{p}_2$, shows the production of R5P from G6P; Figure 11.2.

The two pool vectors that span the left null space are shown in Table 11.2. They represent a balance on glutathione (G_{tot}) and NADP (NP_{tot}), neither of which can leave the system. These are thus the two time-invariant pools associated with the pentose pathway. Both of them are cofactor pools that represent redox potential.

The steady state If we fix the flux into the pathway v_{G6Pin} and the exit flux of R5P, v_{R5P}, then we fix the steady-state flux distribution, since these fluxes uniquely specify the pathway fluxes. We set these numbers at 0.21 mM/h and 0.01 mM/h respectively. The former value represents a typical flux through the pentose pathway, while the latter will couple at a low flux level to the AMP pathways treated in the next chapter. Using $\mathbf{p}_1 + \mathbf{p}_2 = 0.21$ and $\mathbf{p}_2 = 0.01$ we can compute the steady state (last row in Table 11.4).

Table 11.4 The stoichiometric matrix for the pentose pathway. The elemental balancing of the reactions and the two pathway vectors that span the null space are indicated. The steady-state flux (in mm/h) distribution is given in the last row

	I/O				Pentose pathway reactions								Cofactors		Inorganics		
	v_{G6Pin}	v_{F6P}	v_{GAP}	v_{R5P}	v_{G6PDH}	v_{PGLASE}	v_{GL6PDH}	v_{R5PE}	v_{R5PI}	v_{TKI}	v_{TKII}	v_{TALA}	v_{GSSGR}	v_{GSHR}	v_{CO_2}	v_{H}	v_{H_2O}
G6P	1	0	0	0	−1	0	0	0	0	0	0	0	0	0	0	0	0
F6P	0	−1	0	0	0	0	0	0	0	0	1	1	0	0	0	0	0
GAP	0	0	−1	0	0	0	0	0	0	1	1	−1	0	0	0	0	0
GL6P	0	0	0	0	1	−1	0	0	0	0	0	0	0	0	0	0	0
GO6P	0	0	0	0	0	1	−1	0	0	0	0	0	0	0	0	0	0
RU5P	0	0	0	0	0	0	1	−1	−1	0	0	0	0	0	0	0	0
X5P	0	0	0	0	0	0	0	1	0	−1	−1	0	0	0	0	0	0
R5P	0	0	0	−1	0	0	0	0	1	−1	0	0	0	0	0	0	0
S7P	0	0	0	0	0	0	0	0	0	1	0	−1	0	0	0	0	0
E4P	0	0	0	0	0	0	0	0	0	0	−1	1	0	0	0	0	0
NADP	0	0	0	0	−1	0	−1	0	0	0	0	0	1	0	0	0	0
NADPH	0	0	0	0	1	0	1	0	0	0	0	0	−1	0	0	0	0
GSH	0	0	0	0	0	0	0	0	0	0	0	0	2	−2	0	0	0
GSSG	0	0	0	0	0	0	0	0	0	0	0	0	−1	1	0	0	0
CO₂	0	0	0	0	0	0	1	0	0	0	0	0	0	0	−1	0	0
H	0	0	0	0	1	0	1	0	0	0	0	0	−1	2	0	−1	0
H₂O	0	0	0	0	0	−1	0	0	0	0	0	0	0	0	0	0	−1
C	6	−6	−3	−5	0	0	0	0	0	0	0	0	0	0	−1	0	0
H	11	−11	−5	−9	0	0	0	0	0	0	0	0	0	0	0	−1	−2
O	9	−9	−6	−8	0	0	0	0	0	0	0	0	0	0	−2	0	−1
P	1	−1	−1	−1	0	0	0	0	0	0	0	0	0	0	0	0	0
N	0	0	0	0	0	0	0	0	0	0	0	0	0	0	0	0	0
S	0	0	0	0	0	0	0	0	0	0	0	0	0	0	0	0	0
NADP	0	0	0	0	0	0	0	0	0	0	0	0	0	0	0	0	0
P₁	1	$\frac{2}{3}$	$\frac{1}{3}$	0	1	1	1	$\frac{2}{3}$	$\frac{1}{3}$	$\frac{1}{3}$	$\frac{1}{3}$	$\frac{1}{3}$	2	2	1	4	−1
P₂	1	0	0	1	1	1	1	0	1	0	0	0	2	2	1	4	−1
v_stst	0.21	0.13	0.067	0.01	0.21	0.21	0.21	0.13	0.077	0.067	0.067	0.067	0.42	0.42	0.21	0.84	−0.21

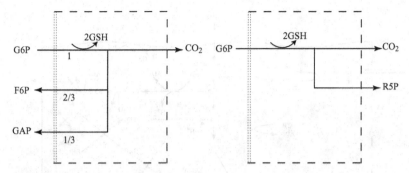

Figure 11.2 A pathway map for the two pathway vectors that span the null space of **S** for the pentose pathway (Table 11.5).

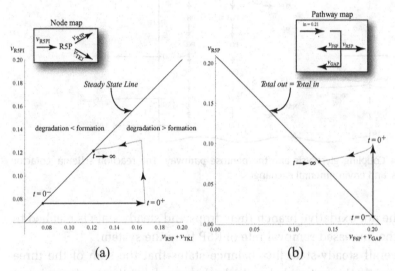

Figure 11.3 Dynamic response of the pentose pathway, increasing the rate of R5P production. (a) The fluxes that form and degrade R5P, $v_{R5P} + v_{TKI}$ versus v_{R5PI}. (b) The fluxes out of the pentose pathway, $v_{F6P} + v_{GAP}$ versus v_{R5P}.

Dynamic response: increased rate of R5P production   One function of the pentose pathway is to provide R5P for biosynthesis. We thus simulate its response to an increased rate of R5P use. We increase the value of k_{R5P} 10-fold at time zero and simulate the response. The responses are best interpreted in terms of the flux balance on the R5P node and the overall flux balance on the pathway; Figure 11.3.

The initial perturbation creates an imbalance on the R5P node. These fluxes must balance in the steady state, as indicated by the 45° line, where the rate of formation and use balance. Initially, v_{R5P} is instantaneously increased $(t = 0^+)$. The immediate response is a compensating increase of production by v_{R5PI}, a rapidly equilibrating enzyme. The utilization of

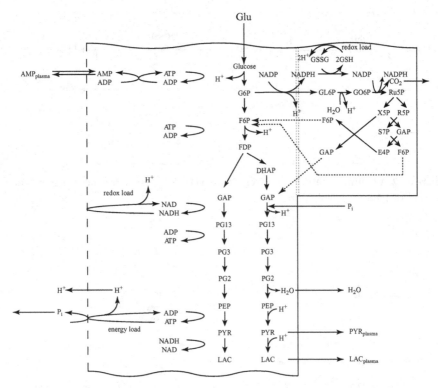

Figure 11.4 Coupling glycolysis and the pentose pathway. The reaction schema, cofactor interactions, and environmental exchanges.

R5P by the nonoxidative branch then drops and steady state is reached to balance the increased removal rate of R5P from the system.

The overall steady-state flux balance states that the sum of the three fluxes leaving the network has to be balanced by the constant input of 0.21 mM/h ($= v_{F6P} + v_{GAP} + v_{R5P}$) as indicated by the $-45°$ line in Figure 11.3b. Initially, v_{R5P} increases 10-fold. Over time, the return flux to glycolysis, $v_{F6P} + v_{GAP}$, decreases, until a steady state is reached.

11.2 The combined stoichiometric matrix

Coupling to glycolysis: forming a unified reaction map Since the inputs to and outputs from the pentose pathway are from glycolysis, the pentose pathway and glycolysis are readily interfaced to form a single reaction map (see Figure 11.4). The dashed arrows in the reaction map represent the return of F6P and GAP from the pentose pathway to glycolysis and do not represent actual reactions. The additional exchanges with the environment, over those in glycolysis alone, are CO_2 secretion and redox load on the NADPH pool, which are shown in Figure 11.4 as a load on GSH. We will not consider R5P production here. It will appear in the next chapter, as it is involved in the nucleotide salvage pathways.

Forming **S** The stoichiometric matrix for glycolysis (Table 10.3) can be appended with the reactions in the pentose pathway (Table 11.3). The resulting stoichiometric matrix is shown in Table 11.5. This matrix has dimensions of 32×32, and its rank is 28. The null and the left null spaces are thus both of dimension 4 ($= 32 - 28$). The matrix is elementally balanced.

Partitioning dashed lines have been put into Table 11.5 to illustrate the structure of the the matrix. The two blocks of matrices on the diagonal are those for each pathway. The lower left block is filled with zero elements showing that the pentose pathway intermediates do not appear in glycolysis. Conversely, the upper right-hand block shows that three of the glycolytic intermediates leave (GAP and F6P) and enter (G6P) the pentose pathway. Both glycolysis and the pentose pathway produce and/or consume protons and water.

The pathway structure: basis for the null space The null space is spanned by four vectors (shown towards the bottom of Table 11.5) that have pathway interpretations (Figure 11.5). The first three pathways are the same as that shown in Figure 10.2 for glycolysis alone, representing p_1, glucose to lactate, p_2, pyruvate to lactate, and p_3, AMP entering and leaving.

A new pathway, p_4, is a cycle through the pentose pathway, where the two F6P output from the pentose pathway flow back up glycolysis through PGI and enter the pentose pathway again, while the GAP output flows down glycolysis and leaves as lactate producing an ATP in lower glycolysis. The net result is the conversion of glucose to three CO_2 molecules and the production of six NADPH redox equivalents. It is a combination of p_1 (Table 11.4) for the pentose pathway alone and the redox neutral use of glycolysis.

This new pathway balances the system fully and is a hybrid of the definitions of the classical glycolytic and pentose pathways. Note that the vectors that span the null space consider the entire network. Thus, as the scope of models increases, the classical pathway definitions give way to network-based pathways that are mathematically defined. This definition is a departure from the historical and heuristic definitions of pathways. This feature is an important one in systems biology.

The time-invariant pools: the basis for the left null space There are four time-invariant pools associated with the coupled glycolytic and pentose pathways (Table 11.7). The first two pools are the same as for glycolysis

Table 11.5 The stoichiometric matrix for the coupled glycolytic and pentose pathways. The null space vectors are also shown. They represent: p_1, the redox neutral glycolysis (glucose to lactate); p_2, the redox exchange with plasma (pyruvate to lactate); p_3, AMP in and out; and p_4, a pentose pathway cycle coupled to glycolysis (the output from the pentose pathway flows up glycolysis and back into the pentose pathway). The steady-state flux solution is given in the last row

	Glycolysis																					Pentose pathway											Σ
	v_{HK}	v_{PGI}	v_{PFK}	v_{TPI}	v_{ALD}	v_{GADPH}	v_{PGK}	v_{PGLM}	v_{ENO}	v_{PK}	v_{LDH}	v_{AMP}	v_{APK}	v_{PYR}	v_{LAC}	v_{ATP}	v_{NADH}	v_{GLUin}	v_{AMPin}	v_{H^+}	v_{H_2O}	v_{G6PDH}	v_{PCLASE}	v_{GL6PDH}	v_{R5PE}	v_{R5PI}	v_{TKI}	v_{TKII}	v_{TALA}	v_{GSSCR}	v_{GSHR}	v_{CO_2}	
GLU	−1	0	0	0	0	0	0	0	0	0	0	0	0	0	0	0	0	1	0	0	0	0	0	0	0	0	0	0	0	0	0	0	2
G6P	1	−1	0	0	0	0	0	0	0	0	0	0	0	0	0	0	0	0	0	0	0	−1	0	0	0	0	0	0	0	0	0	0	3
F6P	0	1	−1	0	0	0	0	0	0	0	0	0	0	0	0	0	0	0	0	0	0	0	0	0	0	0	0	1	1	0	0	0	4
FBP	0	0	1	0	−1	0	0	0	0	0	0	0	0	0	0	0	0	0	0	0	0	0	0	0	0	0	0	0	0	0	0	0	2
DHAP	0	0	0	−1	1	0	0	0	0	0	0	0	0	0	0	0	0	0	0	0	0	0	0	0	0	0	0	0	0	0	0	0	2
GAP	0	0	0	1	1	−1	0	0	0	0	0	0	0	0	0	0	0	0	0	0	0	0	0	0	0	0	1	1	−1	0	0	0	6
PG13	0	0	0	0	0	1	−1	0	0	0	0	0	0	0	0	0	0	0	0	0	0	0	0	0	0	0	0	0	0	0	0	0	2
PG3	0	0	0	0	0	0	1	−1	0	0	0	0	0	0	0	0	0	0	0	0	0	0	0	0	0	0	0	0	0	0	0	0	2
PG2	0	0	0	0	0	0	0	1	−1	0	0	0	0	0	0	0	0	0	0	0	0	0	0	0	0	0	0	0	0	0	0	0	2
PEP	0	0	0	0	0	0	0	0	1	−1	0	0	0	0	0	0	0	0	0	0	0	0	0	0	0	0	0	0	0	0	0	0	3
PYR	0	0	0	0	0	0	0	0	0	1	−1	0	0	−1	0	0	0	0	0	0	0	0	0	0	0	0	0	0	0	0	0	0	2
LAC	0	0	0	0	0	0	0	0	0	0	1	0	0	0	−1	0	0	0	0	0	0	0	0	0	0	0	0	0	0	0	0	0	3
NAD	0	0	0	0	0	−1	0	0	0	0	1	0	0	0	0	0	1	0	0	0	0	0	0	0	0	0	0	0	0	0	0	0	3
NADH	0	0	0	0	0	1	0	0	0	0	−1	0	0	0	0	0	−1	0	0	0	0	0	0	0	0	0	0	0	0	0	0	0	3
AMP	0	0	0	0	0	0	0	0	0	0	0	−1	1	0	0	0	0	0	1	0	0	0	0	0	0	0	0	0	0	0	0	0	6
ADP	1	0	1	0	0	0	−1	0	0	−1	0	1	−2	0	0	1	0	0	0	0	0	0	0	0	0	0	0	0	0	0	0	0	6
ATP	−1	0	−1	0	0	0	1	0	0	1	0	0	1	0	0	−1	0	0	0	0	0	0	0	0	0	0	0	0	0	0	0	0	6

Stoichiometric matrix (metabolites × reactions):

																													v_{sjst}
PHOS	0	0	0	0	0	0	0	0	0	0	0	0	0	0	0	0	0	0	0	0	0	-1	0	0	0	0	0	0		1.12
H	1	-1	1	1	0	0	0	-1	-1	0	0	-1	0	0	0	-1	-1	0	1	0	1	1	2	-1	-1	2	-1		0.91
H_2O	0	0	-1	0	-1	0	0	0	0	0	0	0	0	0	0	0	0	0	0	0	0	0	0	0	0	0	0		1.05
GL6P	0	0	0	0	0	0	0	0	0	0	0	0	0	0	0	0	0	0	0	0	0	0	0	0	0	0	0		1.05
GO6P	0	0	0	0	0	0	0	0	-1	0	0	0	0	0	0	0	0	0	0	0	0	0	0	0	0	0	0		1.05
RU5P	0	0	0	-1	0	-1	-1	1	0	0	0	0	0	0	0	0	0	0	0	0	0	0	0	0	0	0	0		2.17
X5P	0	0	0	1	-1	1	0	0	0	0	0	0	0	0	0	0	0	0	0	0	0	0	0	0	0	0	0		2.17
R5P	0	0	0	0	1	0	1	0	0	0	0	0	0	0	0	0	0	0	0	0	0	0	0	0	0	0	0		2.17
S7P	0	0	0	0	0	-1	0	0	0	0	0	0	0	0	0	0	0	0	0	0	0	0	0	0	0	0	0		2.17
E4P	0	0	0	0	0	1	-1	0	0	0	0	0	0	0	0	0	0	0	0	0	0	0	0	0	0	0	0		2.17
NADP	0	0	-1	0	0	0	0	0	-1	0	0	0	0	0	0	0	0	0	0	0	0	0	0	0	0	0	0		1.95
NADPH	0	0	1	0	0	0	0	0	1	0	0	0	0	0	0	0	0	0	0	0	0	0	0	0	0	0	0		2.17
GSH	0	0	0	0	0	0	0	0	0	0	0	0	0	0	0	0	0	0	0	0	0	0	0	0	0	0	0		0.014
GSSG	0	0	0	0	0	0	0	0	0	0	0	0	0	0	0	0	0	0	0	0	0	0	0	0	0	0	0		0
CO_2	0	0	0	0	0	0	0	0	1	0	0	0	0	0	0	0	0	0	0	0	0	0	0	0	0	0	0		0.224
π_j	5	5	3	4	5	5	2	1	1	1	1	1	1	1	1	13	1	1	5	4	5	2	2	3	2	5	5	2	2	-0.21
P_1	1	1	1	2	2	2	2	0	0	1	0	0	0	2	2	0	2	0	2	2	0	0	0	0	0	0	0	0	0	0.21
P_2	1	0	1	0	0	0	0	-1	0	1	1	-1	0	0	-1	-1	-1	1	0	0	0	0	0	0	0	0	0	0	0	0.21
P_3	0	0	0	0	0	1	1	0	0	0	0	0	0	0	0	0	0	0	3	3	5	2	2	1	1	1	6	6	3	0.14
P_4	1	-2	1	1	1	0	0	1	1	0	0	1	1	0	1	0	1	0	-3	0	0	0	0	0	0	0	0	0	0	0.07
v_{sjst}	1.12	0.91	1.05	1.05	1.05	2.17	2.17	2.17	2.17	2.17	1.95	2.17	0.014	0	0.224	1.12	0.014	3.46	-0.21	0.21	0.21	0.14	0.07	0.07	0.07	0.07	0.42	0.42	0.21	

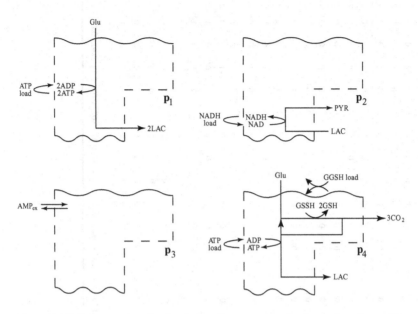

Figure 11.5 Pathway maps for the four pathway vectors for the system formed by coupling glycolysis and the pentose pathway. They span all possible steady-state solutions.

alone: the pool of total NADH and phosphate. Two new pools appear when the pentose pathway is coupled to glycolysis: the total amount of glutathione (G_{tot} in the system),

$$2GSH + GSSG = G_{tot}, \qquad (11.2)$$

and the total amount of the NADPH carrier, NP_{tot},

$$NADPH + NADP = NADPH_{tot} = NP_{tot}. \qquad (11.3)$$

These time-invariant pools are shown in the last four rows of Table 11.7.

11.3 Defining the steady state

Computing the steady-state flux map   The null space is four-dimensional. Thus, if four independent fluxes are specified, then the steady-state flux map is uniquely defined. Therefore, we have to select fluxes for the four steady-state pathway vectors in Table 11.5.

The glucose uptake rate is 1.12 mM/h. The glutathione load is approximately 0.42 mM/h. Thus, we set the weight on $\mathbf{p}_4$ to be $0.42/6 = 0.07$ and that of $\mathbf{p}_1$ to be $1.12 - 0.07 = 1.05$ mM/h. As in Chapter 10, we set the NADH load to be 20% of the glucose uptake; thus, $\mathbf{p}_2$ has a load of $1.12 \times 0.2 = 0.224$ mM/h. Finally, the AMP input rate is the same as before

Table 11.6 Pentose pathway enzymes and transport rates. The numerical value used for ∞ for the equilibrium constants is 10^6

#	Reaction	G	K	Γ	k
1	v_{HK}	0.009	850	0	0.700 007
2	v_{PGI}	0.407	0.41	0.994	2961.11
3	v_{PFK}	0.134	310	0	33.1582
4	v_{TPI}	0.046	0.057	0.796	32.2086
5	v_{ALD}	0.08	0.082	0.973	2657.41
6	v_{GAPDH}	0.007	0.018	0.381	3271.23
7	v_{PGK}	1755	1800	0.975	1.23373×10^6
8	v_{PGLM}	0.146	0.147	0.994	4717.39
9	v_{ENO}	1.504	1.695	0.888	1708.66
10	v_{PK}	19.57	363 000	0	440.186
11	v_{LDH}	44.133	26 300	0.002	1073.94
12	v_{AMP}	0.001	10^6	0	0.161 424
13	v_{ApK}	1.65	1.65	1	1 000 000
14	v_{PYR}	0.995	1	0.995	744.186
15	v_{LAC}	0.735	1	0.735	5.405 56
16	v_{ATP}	0.453	10^6	0	1.356 25
17	v_{NADH}	1.957	10^6	0	7.441 88
18	v_{GLUin}	0	10^6	0	0
19	v_{AMPin}	0	10^6	0	0
20	v_H	1	1	1	100 000
21	v_{H_2O}	1	1	1	100 000
22	v_{G6PDH}	11.875	1000	0.012	21 864.6
23	v_{PGLASE}	21.363	1000	0.021	122.323
24	v_{GL6PDH}	43.341	1000	0.043	29 287.8
25	v_{R5PE}	2.995	3	0.998	16 045.9
26	v_{R5PI}	2.566	2.57	0.999	9664.21
27	v_{TKI}	0.932	1.2	0.777	1675.73
28	v_{TKII}	1.921	10.3	0.187	1146.86
29	v_{TALA}	0.575	1.05	0.548	886.848
30	v_{GSSGR}	0.259	100	0.003	53.3298
31	v_{GSHR}	0.012	2	0.006	0.041 26
32	v_{CO_2}	1	1	1	100 000

at 0.014 mM/h (p_3). Thus, the steady-state flux vector is

$$\mathbf{v}_{stst} = 1.05\mathbf{p}_1 + 0.224\mathbf{p}_2 + 0.014\mathbf{p}_3 + 0.07\mathbf{p}_4. \tag{11.4}$$

This equation is analogous to Eq. (10.3), except the incoming glucose flux is now distributed between the glucose pathway vector ($\mathbf{p}_1$) and the pentose pathway vector ($\mathbf{p}_4$). Summing up the pathway vectors in this ratio gives the steady-state flux values, as shown in the last row of Table 11.5.

Note that this procedure leads to the decomposition of the steady state into four interlinked pathway vectors. All homeostatic states are simultaneously carrying out many functions. This multiplexing can be broken down into underlying pathways and, thus, leads to a clear interpretation of the steady-state solution.

Computing the rate constants   The kinetic constants can be computed from the steady-state values of the concentrations using elementary mass action kinetics. The computation is based on Eq. (10.4). The results from this computation are summarized in Table 11.6. Note that most of the PERCs are large, leading to rapid responses, except for the glutathiones. v_{GSHR} clearly sets the slowest time scale. This table has all the reaction properties that we need to complete the MASS model.

11.4   Simulating the dynamic mass balances

We can now simulate the response of the combined glycolysis and the pentose pathway. We will focus on the responses that we examined for the individual pathways, namely an increase in the rate of ATP and GSH utilization.

Increased rate of ATP utilization   We perform the same simulation as in Chapter 10 by increasing the rate of ATP utilization by 50%. The dynamic phase portraits are shown in Figure 11.6 for the same key fluxes as for glycolysis alone (Figure 10.5). The response is similar, except the dampened oscillations are more pronounced. The oxidative branch of the pentose pathway is not affected, as the GSH load is not changed (Figure 11.6e), while the damped oscillations do occur in the nonoxidative branch (Figure 11.6f).

Increased rate of GSH utilization   The main function of the pentose pathway is to generate redox potential in the form of GSH. We are thus interested in the increased rate of GSH use. We simulate the doubling of the rate of GSH use (see Figure 11.7). The response is characteristic of a highly buffered system. The size of G_{tot} is 3.44 mM ($= 3.2 + 2 \times 0.12$), which is a high concentration for this coupled pathway system.

The glutathione oxidase (the load) and the reductase reach a steady state within an hour at a higher flux level (Figure 11.7a), while the fluxes through the oxidative and nonoxidative branches of the pentose pathway increase, leading to increased CO_2 production. The loss of CO_2 leads to

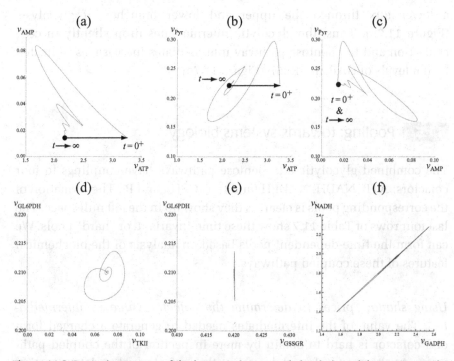

Figure 11.6 Dynamic response of the integrated system of glycolysis and the pentose pathway to increasing the rate of ATP utilization. The dynamic responses of key fluxes are shown. Detailed pairwise phase portraits: (a) v_{ATP} versus v_{AMP}. (b) v_{ATP}, versus v_{Pyr}. (c) v_{AMP} versus v_{Pyr}. (d) v_{TKII} versus v_{GL6PDH}. (e) v_{GSSGR} versus v_{GL6PDH}. (f) v_{GAPDH} versus v_{NADH}. These phase portraits can be compared with the corresponding ones in Figure 10.5. The perturbation is reflected in the instantaneous move of the flux state from the initial steady state to an unsteady state, as indicated by the arrow placing the initial point at $t = 0^+$. The system then returns to its steady state at $t \to \infty$.

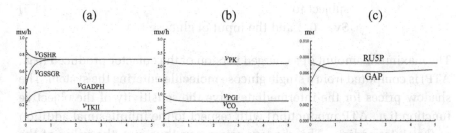

Figure 11.7 Dynamic response of the integrated system of glycolysis and the pentose pathway, doubling the rate of GSH utilization. (a) The dynamic response of the fluxes in the oxidative (v_{G6PDH}) and the nonoxidative (v_{TKII}) branches of the pentose pathway as well as GSH reductase and oxidase (the load). (b) The dynamic response of the upper (v_{PGI}) and lower (v_{PK}) glycolytic fluxes, as well as CO_2 production by the pentose pathway. (c) The dynamic response of sample intermediates in glycolysis, GAP, and the pentose pathway, Ru5P.

a lower flux through the upper and lower branches of glycolysis (Figure 11.7b). Thus, the glycolytic intermediates drop slightly in concentration and the pentose pathway intermediates increase, as indicated by the levels of GAP and Ru5P (Figure 11.7c).

11.5   Pooling: towards systems biology

The combined glycolytic and pentose pathways have couplings to four cofactors: ATP, NADH, NADPH (and thus GSH), and P_i. The formation of the corresponding pools is clear, as they show up in the left null space. The last four rows of Table 11.7 show these time-invariant, or "hard" pools. We can form the time-dependent pools based on analysis of the biochemical features of these coupled pathways.

Using shadow prices to determine the cofactor value of intermediates The value of the intermediates needed to generate a charged form of a cofactor is hard to obtain by mere inspection of the coupled pathways. As the scope of models grows, this cofactor value determination becomes increasingly difficult.

The cofactor values of the intermediates can be obtained using established methods of systems biology. These values can be generated from the shadow prices generated by linear optimization (see [89], Chapter 15). Here, the stoichiometric matrix and the inputs and outputs are used to formulate a linear optimization problem:

$$\begin{aligned} &\max \text{ (cofactor production)}\\ &\text{subject to}\\ &\mathbf{Sv} = \mathbf{0}; \quad \text{and the input of glucose} = 1. \end{aligned} \tag{11.5}$$

The maximum amount of a charged version of the cofactor produced (e.g., ATP) is computed from a single glucose molecule entering the system. The shadow prices for the intermediates give the sensitivity of the objective function (i.e., ATP production) with respect to the infinitesimal addition of that intermediate. The shadow prices can thus give the value of the intermediates for the production of a particular cofactor.

An alternative way to evaluate the cofactor generation value of the intermediates is

$$\begin{aligned} &\max \text{ (cofactor production)}\\ &\text{subject to}\\ &\mathbf{Sv} = \mathbf{0}; \quad \text{and the input of the intermediate of interest} = 1 \end{aligned} \tag{11.6}$$

and by adding an exchange rate for the intermediate of interest. This alternative approach would need many optimization computations, whereas all the shadow prices can be obtained from a single optimization computation.

Defining a pooling matrix   The results from shadow price computations are used in forming the pooling matrix (Table 11.7). The values of the glycolytic intermediates are the same as before, but we can now add the ATP and NADH value of the pentose pathway intermediates. In addition, we can compute the value of the intermediates with respect to generating redox potential in the form of NADPH (or GSH). The high-energy bond value of the intermediates in the combined pathways are shown in the first two lines (GP^+, GP^-). The next two lines do the same for the NADH value (GR^+, GR^-), and then the following two lines give new pools for the NADPH value of the intermediates (GPR^+, GPR^-). The following four pools give the state of the phosphates (P^+, P^-) as well as the energy value of the adenosine phosphates (AP^+, AP^-). These are the same as for glycolysis alone. The next three lines give the redox carriers (N^+, NP^+, G^+). The bottom four pools are the time-invariant pools that are in the left null space of S.

The reactions that move the pools   The fluxes that move the pools can be determined by computing PS. The results are shown in Table 11.8. These pools can be added to the pool-flux map of glycolysis (Figure 10.12).

11.6   Ratios: towards physiology

The five ratios from the glycolytic system alone carry over to the system of the combined pathways. The addition of the pentose pathway adds new property ratios related to the type of redox charge that NADPH carries.

Additional redox charges   The quotient between the charged and uncharged states can be used to define property ratios as in Chapter 10. The ratios are the same here as for glycolysis, with the addition of the following redox charges:

- the NADPH conversion value of the intermediates

$$r_6 = \frac{GRP^+}{GRP^+ + GRP^-};$$
(11.7)

Table 11.7 Definition of functional pools for merged glycolysis and the pentose pathway. The left null space vectors for the stoichiometric matrix for coupled glycolytic and the pentose pathways in Table 11.5 are the last four rows in the table. The four time-invariant pools represent (1) total phosphate, (2) total NADH, (3) total glutathione, and (4) total NADPH

	Glycolysis																				Pentose pathway											
	Glu	G6P	F6P	FBP	DHAP	GAP	PG13	PG3	PG2	PEP	PYR	LAC	NAD	NADH	AMP	ADP	ATP	Phos	H	H₂O	Gl6P	GO6P	Ru5P	X5P	R5P	S7P	E4P	NADP	NADPH	GSH	GSSG	CO₂
GP⁺	2	3	3	4	2	2	2	1	1	1	0	0	0	0	0	0	0	0	0	0	3	3	8/3	8/3	8/3	10/3	7/3	0	0	0	0	0
GP⁻	0	0	0	0	0	0	0	0	0	0	1	1	0	0	0	0	0	0	0	0	0	0	0	0	0	0	0	0	0	0	0	0
GR⁺	2	2	2	2	1	1	1	0	0	0	0	0	0	0	0	0	0	0	0	0	2	2	5/3	5/3	5/3	7/3	4/3	0	0	0	0	0
GR⁻	0	0	0	0	0	0	0	0	0	0	1	1	0	0	0	0	0	0	0	0	0	0	0	0	0	0	0	0	0	0	0	0
GPR⁺	6	6	6	0	0	0	1	1	1	1	1	1	0	0	0	0	0	0	0	0	5	5	4	4	4	8	2	0	0	0	0	0
GPR⁻	0	0	0	1	1	1	1	1	1	0	1	1	0	0	0	0	0	0	0	0	0	0	0	0	0	0	0	0	0	0	0	0
P⁺	0	1	1	2	1	1	1	1	1	1	0	0	0	0	0	1	2	1	0	0	1	1	1	1	1	1	1	0	0	0	0	0
P⁻	0	0	1	0	1	1	1	1	1	1	0	0	0	0	0	0	0	0	0	0	1	1	1	1	1	1	1	0	0	0	0	0
AP⁺	0	0	0	0	0	0	0	0	0	0	0	0	0	0	2	1	2	0	0	0	0	0	0	0	0	0	0	0	0	0	0	0
AP⁻	0	0	0	0	0	0	0	0	0	0	0	0	0	0	2	1	0	0	0	0	0	0	0	0	0	0	0	0	0	0	0	0
N⁺	0	0	0	0	0	0	0	0	0	0	0	0	0	1	0	0	0	0	0	0	0	0	0	0	0	0	0	0	1	0	0	0
NP⁺	0	0	0	0	0	0	0	0	0	0	0	0	0	0	0	0	0	0	0	0	0	0	0	0	0	0	0	0	1	0	0	0
G⁺	0	0	0	0	0	0	0	0	0	0	0	0	0	0	0	0	0	0	0	0	0	0	0	0	0	0	0	0	0	1	0	0
Ptot	0	1	1	2	1	1	2	1	1	1	0	0	0	0	0	1	2	1	0	0	1	1	1	1	1	1	1	0	0	0	0	0
Ntot	0	0	0	0	0	0	0	0	0	0	0	0	1	1	0	0	0	0	0	0	0	0	0	0	0	0	0	0	0	0	0	0
Gtot	0	0	0	0	0	0	0	0	0	0	0	0	0	0	0	0	0	0	0	0	0	0	0	0	0	0	0	0	0	1	2	0
NPtot	0	0	0	0	0	0	0	0	0	0	0	0	0	0	0	0	0	0	0	0	0	0	0	0	0	0	0	1	1	0	0	0

Table 11.8 The fluxes that flow in and out of the pools defined in Table 11.7

Pool name	Fluxes in and out	Size (mM)	Net flux in or out	τ(h)
GP^+	$-v_{GL6PDH}/3 + 2\,v_{GLUin} + v_{HK} + v_{PFK} - v_{PGK} - v_{PK}$	3.000 11	4.41	0.680
GP^-	$-v_{LAC} + v_{PK} - v_{PYR}$	1.420 3	2.17	0.655
GR^+	$-v_{GAPDH} - v_{GL6PDH}/3 + 2v_{GLUin} - v_{LAC} + v_{LDH}$	3.888 46	4.186	0.929
GR^-	$v_{GAPDH} - v_{LDH} - v_{PYR}$	0.166 144	2.17	0.077
GRP^+	$-v_{G6PDH} - v_{GL6PDH} + 6v_{GLUin} - 6v_{PFK}$	6.938 16	6.72	1.03
GRP^-	$v_{ALD} - v_{LAC} + v_{PFK} - v_{PYR} - v_{TALA} + v_{TKI} + v_{TKII}$	1.708 02	2.24	0.763
P^+	$-v_{ATP} - v_{G6PDH} + v_{PK} + v_{TKI} + 2v_{TKII}$	3.755 12	2.38	1.578
P^-	$v_{GAPDH} - v_{PK}$	0.105 843	2.17	0.049
AP^+	$-v_{ATP} - v_{HK} - v_{PFK} + v_{PGK} + v_{PK}$	3.49	4.34	0.804
AP^-	$-2\,v_{AMP} + 2v_{AMPin} + v_{ATP} + v_{HK} + v_{PFK} - v_{PGK} - v_{PK}$	0.463 456	4.368	0.106
N^+	$v_{GAPDH} - v_{LDH} - v_{NADH}$	0.030 1	2.17	0.014
NP^+	$v_{G6PDH} + v_{GL6PDH} - v_{GSSGR}$	0.065 8	0.42	0.157
G^+	$2\,v_{GSSGR} - 2v_{GSHR}$	3.2	0.84	3.810
P_{tot}	0	6.461 65		∞
N_{tot}	0	0.089		∞
G_{tot}	0	3.44		∞
NP_{tot}	0	0.066		∞

- the NADPH carrier

$$r_7 = \frac{\text{NADPH}}{\text{NADPH} + \text{NADP}} = \frac{NP^+}{NP^+ + NP^-}; \qquad (11.8)$$

- the glutathione carrier

$$r_8 = \frac{\text{GSH}}{\text{GSH} + 2\text{GSSG}}. \qquad (11.9)$$

Dynamic responses of the ratios The response to the increased rate of GSH utilization on the cofactor charge ratios is shown in Figure 11.8. The glutathione redox charge drops from 0.93 to 0.88 (Figure 11.8a) and complementary to that there is a modest increase in the NADPH redox charge (Figure 11.8b). Figure 11.8c and d show the dip in the NADH redox charge and the adenosine phosphate energy charge that result from the effect that increased pentose pathway flux has on reduced glycolysis flux. These ratios drop modestly but then return almost to their exact initial value. Thus, there is dynamic interaction between the pathways, but the steady state remains similar.

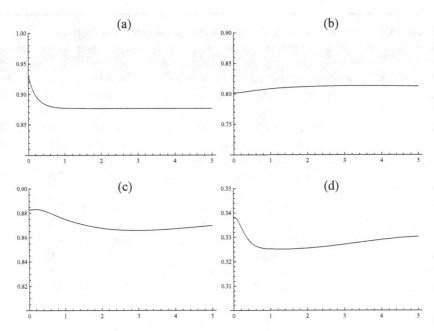

Figure 11.8 The dynamic response of the combined glycolytic and pentose phosphate path-way ratios to a doubling of rate of GSH utilization; see Figure 11.7. (a) The response of the glutathione redox ratio. (b) The response of the NADPH redox ratio. (c) The response of the adenosine phosphate energy charge. (d) The response of the NADH redox ratio.

11.7   Summary

➤ A stoichiometric matrix for the pentose pathway can be formed and a MASS model built to simulate the dynamic responses of this pathway.

➤ The stoichiometric matrix for the pentose pathway can be merged with the stoichiometric matrix for glycolysis to form a system that describes the coupling of the two pathways.

➤ By going through the process of coupling the pentose pathway to glycolysis, we begin to see the emergence of systems biology with scale.

➤ The basis for the null space of **S** gets complicated. The primary pentose pathway vector now becomes cyclic, as it connects the glycolytic inputs and outputs from the pentose pathway. The pentose pathway thus integrates with glycolysis.

➤ To define the composition of the redox pools we have to deploy linear programming, as it becomes difficult to determine the redox value of the intermediates by simple inspection. This difficulty arises from the coupling of two processes and the growing scale of the model. Linear programming optimization can solve this problem.

➤ The response of the coupled pathways to an increase to the rate of use of the redox potential generated by the pentose pathway shows the importance of GSSG/GSH buffering. The concentration of glutathione is high and buffers the response.

➤ Nevertheless, the pathways are coupled and this perturbation does influence glycolysis. The simulations show the interactions between glycolysis and the pentose pathway when we put energy and redox loads on the coupled pathways. Thus, it may be detrimental to think of the pentose pathway as just the producer of R5P, as one needs many of the glycolytic reactions to form R5P.

➤ Therefore, we have to begin to think about the network as a whole. This point of view becomes even more apparent in the next chapter.

## CHAPTER 12

Building networks

Chapter 11 described the integration of two pathways. We will now add yet another set of reactions to the system. The AMP input and output that was described earlier by two simple reactions across the system boundary represent a network of reactions that are called *degradation and salvage pathways*. The salvage pathways recycle the nucleotide by reversing their degradation pathways. AMP degradation and salvage are low-flux pathways but have genetic defects that cause serious diseases. Thus, even though fluxes through these pathways are low, their function is critical. As we have seen, many of the long-term dynamic responses are determined by the adjustment of the total adenylate cofactor pool. In this chapter, we integrate the degradation and salvage pathways to form the core metabolic network in the red blood cell.

12.1 AMP metabolism

AMP metabolism forms a small sub-network in the overall red blood cell metabolic network. We will first consider its properties before we integrate it with the combined glycolytic and pentose pathway model.

The AMP metabolic sub-network AMP is simultaneously degraded and synthesized, creating a dynamic balance (Figure 12.1). The metabolites that are found in these pathways and that are over and above those that are in the integrated glycolytic and pentose pathway network are listed in Table 12.1. To build this sub-network, prior to integration with the glycolytic and pentose pathway network, we need consider a sub-network comprised of the metabolites shown in Table 12.2. The reactions that take place in the degradation and biosynthesis of AMP are shown in Table 12.3.

Table 12.1 Intermediates of AMP degradation and synthesis, their abbreviations, and steady-state concentrations. The concentrations given are those typical for the human red blood cell. The index on the compounds is added to that for the combined glycolysis and pentose pathway (Table 11.1)

#	Abbreviation	Intermediates	Concentration (mM)
33	ADO	Adenosine	0.0012
34	ADE	Adenine	0.001
35	IMP	Inosine monophosphate	0.01
36	INO	Inosine	0.001
37	HYP	Hypoxanthine	0.002
38	R1P	Ribose 1-phosphate	0.06
39	PRPP	Phosphoribosyl diphosphate	0.005

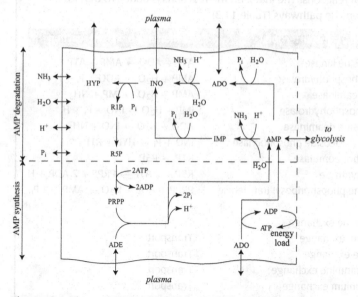

Figure 12.1 The nucleotide metabolism considered in this chapter. There are two biochemical ways in which AMP is degraded and two ways in which it is synthesized.

Degradation   AMP can be degraded either by dephosphorylation or by deamination. In the former case, adenosine (ADO) is formed and can cross the cell membrane to enter plasma. In the latter case, IMP is formed and can then subsequently be dephosphorylated to form inosine (INO). INO can then cross the cell membrane, or be further degraded to form a pentose (R1P) and the purine base hypoxanthine (HYP), the latter of which can be exchanged with plasma. HYP will become uric acid, the accumulation of which causes hyperuricemia (gout).

Table 12.2 Elemental composition of the intermediates of the AMP metabolic sub-network. The compounds in the block to the right are new and described in Table 12.1

	AMP	ADP	ATP	P_i	H	H_2O	NH_3	ADO	ADE	IMP	INO	HYP	R1P	R5P	PRPP
C	10	10	10	0	0	0	0	10	5	10	10	5	5	5	5
H	13	13	13	1	1	2	3	13	5	12	12	4	9	9	8
O	7	10	13	4	0	1	0	4	0	8	5	1	8	8	14
P	1	2	3	1	0	0	0	0	0	1	0	0	1	1	3
N	5	5	5	0	0	0	1	5	5	4	4	4	0	0	0

Table 12.3 AMP degradation and biosynthetic pathway enzymes and transporters, their abbreviations, and chemical reactions. The index on the reactions is added to that for the combined glycolysis and pentose pathways (Table 11.3)

#	Abbrev.	Enzymes/transporter/load	Elementally balanced reaction
33	AK	Adenosine kinase	$ATP + ADO \rightarrow AMP + ATP$
34	AMPase	AMP phosphohydrolase	$AMP + H_2O \rightarrow ADO + P_i + H$
35	AMPDA	AMP deaminase	$AMP + H_2O \rightarrow IMP + NH_3$
36	IMPase	IMP phosphohydrolase	$IMP + H_2O \rightarrow INO + P_i + H$
37	ADA	Adenosine deaminase	$ADO + H_2O \rightarrow INO + NH_3$
38	PNPase	Purine nucleoside phosphorylase	$INO + P_i \rightarrow HYP + R1P$
39	PRM	Phosphoribomutase	$R1P \rightarrow R5P$
40	PRPPsyn	PRPP synthase	$R5P + 2 ATP \rightarrow PRPP + 2 ADP + H$
41	ADPRT	Adenine phopshoribosyl transferase	$PRPP + ADE + H_2O \rightarrow AMP + 2 P_i$
42	ADO	Adenosine exchange	Transport
43	ADE	Adenine exchange	Transport
44	INO	Inosine exchange	Transport
45	HYP	Hypoxanthine exchange	Transport
46	NH_3	Ammonium exchange	Transport

Biosynthesis The red blood cell does not have the capacity for *de novo* adenine synthesis. It can synthesize AMP in two different ways. First, it can phosphorylate ADO directly to form AMP using an ATP to ADP conversion. This ATP use creates an energy load met by glycolysis. A second process to offset degradation of adenine is the salvage pathway, in which adenine (ADE) is picked up from plasma and is combined with phosphoribosyl diphosphate (PRPP) to form AMP. PRPP is formed from R5P using two ATP equivalents, where the R5P is formed from isomerization of R1P that is formed during INO degradation. In this way, the pentose is recycled to resynthesize AMP at the cost of two ATP molecules. The R5P can also come from the pentose phosphate pathway in an integrated network.

Stoichiometric matrix for AMP metabolism   The properties of the stoichiometric matrix are shown in Table 12.4. All the reactions are elementally balanced except for the exchange reactions. The matrix has dimensions of 15×19 and a rank of 14. It thus has a five-dimensional null space and a one-dimensional left null space.

The pathway vectors   Five pathway vectors chosen to span the null space are shown towards the bottom of Table 12.4. They can be divided into groups of degradative and biosynthetic pathways.

The first three pathways, $\mathbf{p}_1$, $\mathbf{p}_2$, and $\mathbf{p}_3$, are degradative pathways of AMP. The first pathway degrades AMP to ADO via dephosphorylation. The second pathway degrades AMP to IMP via AMP deaminase, followed by dephosphorylation and secretion of INO. The third pathway degrades AMP, first with dephosphorylation to ADO followed by deamination to INO and its secretion. These are shown graphically on the reaction map in Figure 12.2. Note that the second and third pathways are equivalent overall, but take an alternative route through the network. These two pathways are said to have an equivalent input/output signature [89].

The fourth pathway, $\mathbf{p}_4$, shows the import of adenosine (ADO) and its phosphorylation via direct use of ATP. Thus, the cost of the AMP molecule now, relative to the plasma environment, is one high-energy phosphate bond. In the previous chapters, AMP was directly imported and did not cost any high-energy bonds in the defined system. Pathway four is essentially the opposite of pathway one, except it requires ATP for fuel. If one sums these two pathways, a futile cycle results.

The fifth pathway, $\mathbf{p}_5$, is a salvage pathway. It takes the INO produced through degradation of AMP, cleaves the pentose off to form and secrete hypoxanthine (HYP), and recycles the pentose through the formation of PRPP at the cost of two ATP molecules. PPRP is then combined with an imported adenine base to form AMP. This pathway is energy-requiring, needing two ATPs. As we detail below, some of the chemical reactions of this pathway change when it is integrated with glycolysis and the pentose pathway, as we have modified the PRPP synthase reaction to make it easier to analyze this sub-network.

The pool vector   The left null space is one-dimensional. It has one time-invariant: ATP + ADP. In this sub-network, this sum acts like a conserved cofactor.

Table 12.4 The stoichiometric matrix for AMP degradation and biosynthetic pathways. The elemental balancing of the reactions is shown. The reactions to the right of the vertical dashed line are transporters that will not be elementally balanced. The five pathway vectors that span the null space are indicated. The steady-state flux distribution is given in the last row

	Nucleotide metabolism										Exchange fluxes								
	v_{AK}	v_{AMPase}	v_{AMPDA}	v_{ImPase}	v_{ADA}	v_{PpPase}	v_{PRM}	v_{ATPgen}	$v_{PRPPsyn}$	v_{ADPRT}	v_{ADO}	v_{ADE}	v_{INO}	v_{HYP}	v_{AMP}	v_{H}	v_{H_2O}	v_{P_i}	v_{NH_3}
AMP	1	−1	−1	0	0	0	0	0	0	1	0	0	0	0	−1	0	0	0	0
ADP	1	0	0	0	0	0	0	−1	2	0	0	0	0	0	0	0	0	0	0
ATP	−1	0	0	0	0	0	0	1	−2	0	0	0	0	0	0	0	0	0	0
P_i	0	1	0	1	0	−1	0	−1	0	2	0	0	0	0	0	0	0	−1	0
H	0	0	−1	0	−1	0	0	1	1	0	0	0	0	0	0	−1	0	0	0
H_2O	0	−1	−1	−1	−1	0	0	0	0	0	0	0	0	0	0	0	−1	0	0
NH_3	0	0	1	0	1	0	0	0	0	0	0	0	0	0	0	0	0	0	−1
ADO	0	1	0	0	−1	0	0	0	0	0	−1	0	0	0	0	0	0	0	0
ADE	0	0	0	0	0	0	0	0	0	−1	0	−1	0	0	0	0	0	0	0
IMP	0	0	1	−1	0	0	0	0	0	0	0	0	0	0	0	0	0	0	0
INO	0	0	0	1	1	−1	0	0	0	0	0	0	−1	0	0	0	0	0	0
HYP	0	0	0	0	0	1	0	0	0	0	0	0	0	−1	0	0	0	0	0
R1P	0	0	0	0	0	1	−1	0	0	0	0	0	0	0	0	0	0	0	0
R5P	0	0	0	0	0	0	1	0	−1	0	0	0	0	0	0	0	0	0	0
PRPP	0	0	0	0	0	0	0	0	1	−1	0	0	0	0	0	0	0	0	0
C	0	0	0	0	0	0	0	0	0	0	−10	−5	−10	−5	−10	0	0	0	0
H	0	0	0	0	0	0	0	0	0	0	−13	−5	−12	−4	−13	−1	−2	−1	−3
O	0	0	0	0	0	0	0	0	0	0	−4	0	−5	−1	−7	0	−1	−4	0
P	0	0	0	0	0	0	0	0	0	0	0	0	0	0	−1	0	0	−1	0
N	0	0	0	0	0	0	0	0	0	0	−5	−5	−4	−4	−5	0	0	0	−1
P_1	0	1	0	0	0	0	0	0	0	0	1	0	0	0	−1	−1	−1	−1	0
P_2	0	0	1	1	0	0	0	0	0	0	0	0	1	0	−1	1	−2	−1	1
P_3	0	0	0	0	1	0	0	0	0	0	1	0	1	0	−1	1	−2	−1	1
P_4	1	0	0	0	0	0	0	−1	0	0	−1	−1	0	0	0	−1	0	−1	0
P_5	0	0	0	0	0	1	1	0	0	1	0	−1	0	1	−1	0	−1	0	1
v_{sst}	0.12	0.12	0.014	0.014	0.01	0.014	0.014	0.15	0.014	0.014	−0.01	−0.014	0.01	0.014	0.0	0.0	−0.024	0.0	0.024

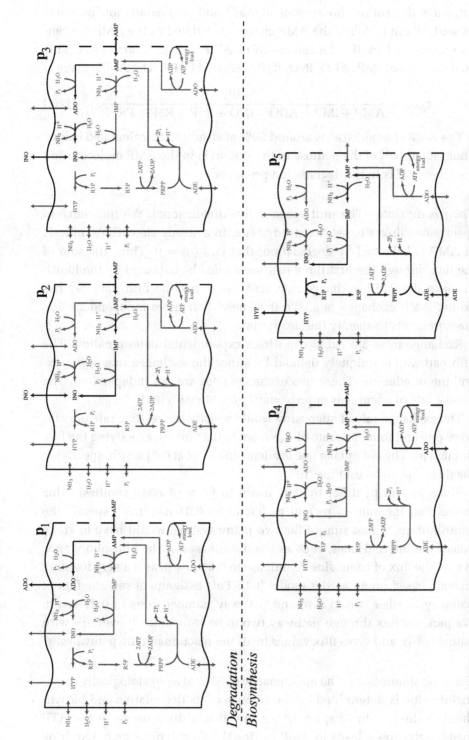

Figure 12.2 The graphical depiction of the five chosen pathway vectors of the null space of the stoichiometric matrix, shown in Table 12.4. The first three pathways are degradative pathways, while the fourth is a direct synthesis pathway and the fifth is a salvage pathway.

Definition of pools and ratios   The AMP degradation and biosynthetic pathways determine the amount of AMP and its degradation products present. We thus define the AMP charge, calculated as the AMP concentration divided by the concentration of AMP and its degradation. This pool consists of IMP, ADO, INO, R1P, R5P, and PRPP. The AMP charge is

$$r_{AMP} = \frac{AMP}{AMP + IMP + ADO + INO + R1P + R5P + PRPP}. \quad (12.1)$$

The AMP charge, r_{AMP}, is around 50% at steady state, calculated below. Thus, about half of the pentose in the system is in the AMP molecule and the other half is in the degradation products.

The steady state   The null space is five-dimensional. We thus have to specify five fluxes to set the steady state. In a steady state, the synthesis of AMP is balanced by degradation; that is, $v_{AMP} = 0$. Thus, the sum of the flux through the first three pathways must be balanced by the fourth to make the AMP exchange rate zero. Note that the fifth pathway has no net AMP exchange rate. We then need four more independent flux measurements to specify the steady state.

Exchange rates are subject to direct experimental determination. The fifth pathway is uniquely defined by either the exchange rate of hypoxanthine or adenine. These two exchange rates are not independent. The uptake rate of adenine is approximately 0.014 mM/h [62].

The exchange rate of adenosine would specify the relative rate of pathways one and four. The rate of v_{AK} is set to 0.12 mM/h, specifying the flux through p_4. The net uptake rate of adenosine is set at 0.01 mM/h, specifying the flux of p_1 to be 0.11 mM/h.

Since p_1 and p_4 differ by 0.01 mM/h in favor of AMP synthesis, this means that the sum of p_2 and p_3 has to be 0.01 mM/h. To specify the contributions to that sum of the two pathways, we would have to know one of the internal rates, such as the deaminases or the phosphorylases. We set the flux of adenosine deaminase to 0.01 mM/h, as it a very low-flux enzyme based on an earlier model [62]. This assignment sets the flux of pathway number two to zero and pathway number three to 0.01 mM/h. We pick the flux through pathway two to be zero, since it overlaps with pathway five and gives flux values to all the reactions in the pathways.

Dynamic simulation   The maintenance of AMP at a physiologically meaningful value is determined by the balance of its degradative and biosynthetic pathways. In Chapter 10 we saw that a disturbance in the ATP load on glycolysis leads to AMP exiting the glycolytic system, requiring degradation. Here, we are interested in seeing how the balance of the AMP pathways is influenced by a sudden change in the level of AMP.

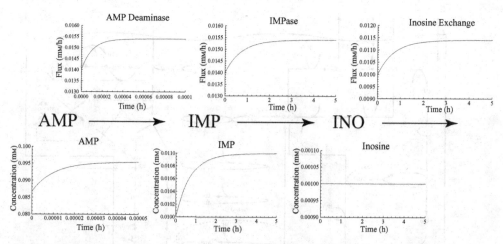

Figure 12.3 The dynamic response of the AMP synthesis/degradation sub-network to a sudden 10% increase in the AMP concentration at $t = 0$. The most notable set of fluxes are shown.

We thus simulate the response to a sudden 10% increase in AMP concentration. The response of the degradative pathways is a rapid conversion of AMP to IMP followed by a slow conversion of IMP to INO that is then rapidly exchanged with plasma; Figure 12.3. This degradation route is kinetically preferred. A sudden decrease in AMP has similar, but opposite, responses. If AMP is suddenly reduced in concentration, the flux through this pathway drops.

12.2 Network integration

The AMP metabolic sub-network of Figure 12.1 can be integrated with the glycolysis and pentose pathway network from Chapter 11. The result is shown in Figure 12.4. This integrated network represents the core of the metabolic network in the red blood cell, the simplest cell in the human body.

Integration issues Given the many points of contact created between the AMP sub-network and the combined glycolytic and pentose pathway network, there are a few interesting integration issues. They are highlighted with dashed ovals in Figure 12.4 and are as follows:

- The AMP molecule in the two networks connects the two. These two nodes need to be merged into one.
- The R5P molecule appears in both the AMP metabolic sub-network and the pentose pathway, so these two nodes also need to be merged.

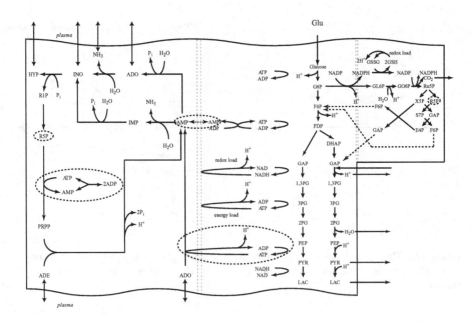

Figure 12.4 The core metabolic network in the human red blood cell comprised of glycolysis, the pentose pathway, and adenine nucleotide metabolism. Some of the integration issues discussed in the text are highlighted with dashed ovals.

- In the sub-network described above, the stoichiometry of the PRPP synthase reaction is

$$R5P + 2ATP \rightarrow PRPP + 2ADP + H, \tag{12.2}$$

but in actuality it is

$$R5P + ATP \rightarrow PRPP + AMP + H; \tag{12.3}$$

this difference disappears, since the ApK reaction is in the combined glycolytic and pentose pathway network:

$$AMP + ATP \rightleftharpoons 2ADP; \tag{12.4}$$

i.e., if Equations (12.3) and (12.4) are added, one gets Eq. (12.2).
- The ATP cost of driving the biosynthetic pathways to AMP is now a part of the ATP load in the integrated model.
- The AMP exchange reaction disappears, and instead we now have exchange reactions for ADO, ADE, INO, and HYP, and the deamination reactions create an exchange flux for NH_3.

The stoichiometric matrix These integration issues can be incorporated into the reaction list and the stoichiometric matrix can be formulated; Table 12.5. There are 40 compounds and 45 reactions in this network. The

Table 12.5 The stoichiometric matrix for the coupled glycolytic, pentose, and AMP pathways. The matrix is organized to order of the sub-networks as follows: glycolysis, pentose pathway, AMP metabolism, cofactors, glycolytic metabolite exchanges, metabolite exchanges in AMP metabolism, and exchanges of inorganic compounds. The eight pathways that span the null space are shown in the final eight rows

Column groups (left to right): **Glycolysis** (v_{HK}, v_{PGI}, v_{PFK}, v_{TPI}, v_{ALD}, v_{GAPDH}, v_{PGK}, v_{PGLM}, v_{ENO}, v_{PK}, v_{LDH}); **Pentose pathway** (v_{G6PDH}, v_{PGLase}, v_{GL6PDH}, v_{R5PE}, v_{R5PI}, v_{TKI}, v_{TKII}, v_{TALA}); **AMP metabolism** (v_{AK}, v_{AMPase}, v_{ADA}, v_{AMPDA}, v_{IMPASE}, v_{PNPASE}, v_{PRM}, $v_{PRPPSYN}$, v_{ADPRT}, v_{Apk}); **Cofactors** (v_{ATP}, v_{NADH}, v_{GSSGR}, v_{GSHR}); **Glycolytic exchanges** (v_{GLUin}, v_{PYR}, v_{LAC}); **AMP metabolic exchanges** (v_{ADE}, v_{ADO}, v_{INO}, v_{HYP}); **Inorganics** (v_{PHOS}, v_{H}, v_{H2O}, v_{CO2}, v_{NH3}).

	v_{HK}	v_{PGI}	v_{PFK}	v_{TPI}	v_{ALD}	v_{GAPDH}	v_{PGK}	v_{PGLM}	v_{ENO}	v_{PK}	v_{LDH}	v_{G6PDH}	v_{PGLase}	v_{GL6PDH}	v_{R5PE}	v_{R5PI}	v_{TKI}	v_{TKII}	v_{TALA}	v_{AK}	v_{AMPase}	v_{ADA}	v_{AMPDA}	v_{IMPASE}	v_{PNPASE}	v_{PRM}	$v_{PRPPSYN}$	v_{ADPRT}	v_{Apk}	v_{ATP}	v_{NADH}	v_{GSSGR}	v_{GSHR}	v_{GLUin}	v_{PYR}	v_{LAC}	v_{ADE}	v_{ADO}	v_{INO}	v_{HYP}	v_{PHOS}	v_{H}	v_{H2O}	v_{CO2}	v_{NH3}
GLU	-1	0	0	0	0	0	0	0	0	0	0	0	0	0	0	0	0	0	0	0	0	0	0	0	0	0	0	0	0	0	0	0	0	1	0	0	0	0	0	0	0	0	0	0	0
G6P	1	-1	0	0	0	0	0	0	0	0	0	-1	0	0	0	0	0	0	0	0	0	0	0	0	0	0	0	0	0	0	0	0	0	0	0	0	0	0	0	0	0	0	0	0	0
F6P	0	1	-1	0	0	0	0	0	0	0	0	0	0	0	0	0	0	1	1	0	0	0	0	0	0	0	0	0	0	0	0	0	0	0	0	0	0	0	0	0	0	0	0	0	0
FBP	0	0	1	0	-1	0	0	0	0	0	0	0	0	0	0	0	0	0	0	0	0	0	0	0	0	0	0	0	0	0	0	0	0	0	0	0	0	0	0	0	0	0	0	0	0
DHAP	0	0	0	-1	1	0	0	0	0	0	0	0	0	0	0	0	0	0	0	0	0	0	0	0	0	0	0	0	0	0	0	0	0	0	0	0	0	0	0	0	0	0	0	0	0
GAP	0	0	0	1	1	-1	0	0	0	0	0	0	0	0	0	0	1	1	-1	0	0	0	0	0	0	0	0	0	0	0	0	0	0	0	0	0	0	0	0	0	0	0	0	0	0
PG13	0	0	0	0	0	1	-1	0	0	0	0	0	0	0	0	0	0	0	0	0	0	0	0	0	0	0	0	0	0	0	0	0	0	0	0	0	0	0	0	0	0	0	0	0	0
PG3	0	0	0	0	0	0	1	-1	0	0	0	0	0	0	0	0	0	0	0	0	0	0	0	0	0	0	0	0	0	0	0	0	0	0	0	0	0	0	0	0	0	0	0	0	0
PG2	0	0	0	0	0	0	0	1	-1	0	0	0	0	0	0	0	0	0	0	0	0	0	0	0	0	0	0	0	0	0	0	0	0	0	0	0	0	0	0	0	0	0	0	0	0
PEP	0	0	0	0	0	0	0	0	1	-1	0	0	0	0	0	0	0	0	0	0	0	0	0	0	0	0	0	0	0	0	0	0	0	0	0	0	0	0	0	0	0	0	0	0	0
PYR	0	0	0	0	0	0	0	0	0	1	-1	0	0	0	0	0	0	0	0	0	0	0	0	0	0	0	0	0	0	0	0	0	0	0	-1	0	0	0	0	0	0	0	0	0	0
LAC	0	0	0	0	0	0	0	0	0	0	1	0	0	0	0	0	0	0	0	0	0	0	0	0	0	0	0	0	0	0	0	0	0	0	0	-1	0	0	0	0	0	0	0	0	0
GL6P	0	0	0	0	0	0	0	0	0	0	0	1	-1	0	0	0	0	0	0	0	0	0	0	0	0	0	0	0	0	0	0	0	0	0	0	0	0	0	0	0	0	0	0	0	0
GO6P	0	0	0	0	0	0	0	0	0	0	0	0	1	-1	0	0	0	0	0	0	0	0	0	0	0	0	0	0	0	0	0	0	0	0	0	0	0	0	0	0	0	0	0	0	0
RU5P	0	0	0	0	0	0	0	0	0	0	0	0	0	1	-1	-1	0	0	0	0	0	0	0	0	0	0	0	0	0	0	0	0	0	0	0	0	0	0	0	0	0	0	0	0	0
X5P	0	0	0	0	0	0	0	0	0	0	0	0	0	0	1	0	-1	-1	0	0	0	0	0	0	0	0	0	0	0	0	0	0	0	0	0	0	0	0	0	0	0	0	0	0	0
R5P	0	0	0	0	0	0	0	0	0	0	0	0	0	0	0	1	-1	0	0	0	0	0	0	0	0	1	-1	0	0	0	0	0	0	0	0	0	0	0	0	0	0	0	0	0	0
S7P	0	0	0	0	0	0	0	0	0	0	0	0	0	0	0	0	1	0	-1	0	0	0	0	0	0	0	0	0	0	0	0	0	0	0	0	0	0	0	0	0	0	0	0	0	0
E4P	0	0	0	0	0	0	0	0	0	0	0	0	0	0	0	0	0	-1	1	0	0	0	0	0	0	0	0	0	0	0	0	0	0	0	0	0	0	0	0	0	0	0	0	0	0
ADE	0	0	0	0	0	0	0	0	0	0	0	0	0	0	0	0	0	0	0	0	0	0	0	0	0	0	0	-1	0	0	0	0	0	0	0	0	1	0	0	0	0	0	0	0	0
ADO	0	0	0	0	0	0	0	0	0	0	0	0	0	0	0	0	0	0	0	0	1	-1	0	0	0	0	0	0	-1	0	0	0	0	0	0	0	0	1	0	0	0	0	0	0	0
IMP	0	0	0	0	0	0	0	0	0	0	0	0	0	0	0	0	0	0	0	0	0	0	1	-1	0	0	0	0	0	0	0	0	0	0	0	0	0	0	0	0	0	0	0	0	0
INO	0	0	0	0	0	0	0	0	0	0	0	0	0	0	0	0	0	0	0	0	0	1	0	1	-1	0	0	0	0	0	0	0	0	0	0	0	0	0	1	0	0	0	0	0	0
HYP	0	0	0	0	0	0	0	0	0	0	0	0	0	0	0	0	0	0	0	0	0	0	0	0	1	0	0	0	0	0	0	0	0	0	0	0	0	0	0	1	0	0	0	0	0

(continued)

Table 12.5 (continued).

	Glycolysis											Pentose pathway								AMP metabolism										Cofactors				Glycolytic exchanges			AMP metabolic exchanges				Inorganics				
	VHK	VPGI	VPFK	VTPI	VALD	VGAPDH	VPGK	VPGLM	VENO	VPK	VLDH	VG6PDH	VPGLase	VGL6PDH	VRSPE	VRSPI	VTKI	VTKII	VTALA	VAK	VAMPase	VADA	VAMPDA	VIMPASE	VPNPASE	VPRM	VPRPPSYN	VADPRT	VApk	VATP	VNADH	VGSSGR	VGSHR	VGLUin	VPYR	VLAC	VADE	VADO	VINO	VHYP	VPHOS	VH	VH2O	VCO2	VNH3
R1P	0	0	0	0	0	0	0	0	0	0	0	0	0	0	0	0	0	0	0	0	0	0	0	0	1	-1	0	0	0	0	0	0	0	0	0	0	0	0	0	0	0	0	0	0	0
PRPP	0	0	0	0	0	0	0	0	0	0	0	0	0	0	0	0	0	0	0	0	0	0	0	0	0	0	1	-1	-1	0	0	0	0	0	0	0	0	0	0	0	0	0	0	0	0
NAD	0	0	0	0	0	-1	0	0	0	0	1	0	0	0	0	0	0	0	0	0	0	0	0	0	0	0	0	0	0	0	-1	0	0	0	0	0	0	0	0	0	0	0	0	0	0
NADH	0	0	0	0	0	1	0	0	0	0	-1	0	0	0	0	0	0	0	0	0	0	0	0	0	0	0	0	0	0	0	1	0	0	0	0	0	0	0	0	0	0	0	0	0	0
AMP	0	0	0	0	0	0	0	0	0	0	0	0	0	0	0	0	0	0	0	-1	-1	0	-1	0	0	0	1	1	1	0	0	0	0	0	0	0	0	0	0	0	0	0	0	0	0
ADP	1	0	1	0	0	0	-1	0	0	-1	0	0	0	0	0	0	0	0	0	2	0	0	0	0	0	0	0	0	1	1	0	0	0	0	0	0	0	0	0	0	0	0	0	0	0
ATP	-1	0	-1	0	0	0	1	0	0	1	0	0	0	0	0	0	0	0	0	-1	0	0	0	0	0	0	-1	0	-1	-1	0	0	0	0	0	0	0	0	0	0	0	0	0	0	0
NADP	0	0	0	0	0	0	0	0	0	0	0	-1	0	-1	0	0	0	0	0	0	0	0	0	0	0	0	0	0	0	0	0	1	0	0	0	0	0	0	0	0	0	0	0	0	0
NADPH	0	0	0	0	0	0	0	0	0	0	0	1	0	1	0	0	0	0	0	0	0	0	0	0	0	0	0	0	0	0	0	-1	0	0	0	0	0	0	0	0	0	0	0	0	0
GSH	0	0	0	0	0	0	0	0	0	0	0	0	0	0	0	0	0	0	0	0	0	0	0	0	0	0	0	0	0	0	0	2	-2	0	0	0	0	0	0	0	0	0	0	0	0
GSSG	0	0	0	0	0	0	0	0	0	0	0	0	0	0	0	0	0	0	0	0	0	0	0	0	0	0	0	0	0	0	0	-1	1	0	0	0	0	0	0	0	0	0	0	0	0
Pi	0	0	0	0	0	-1	0	0	0	0	0	0	0	0	0	0	0	0	0	0	1	0	0	1	-1	0	0	0	0	1	0	0	0	0	0	0	0	0	0	0	-1	0	0	0	0
H	1	0	0	0	0	1	0	0	0	0	-1	1	0	1	0	0	0	0	0	0	0	0	0	0	0	0	0	0	0	1	0	0	0	0	0	0	0	0	0	0	0	-1	0	0	0
H2O	0	0	0	0	0	0	0	0	0	0	0	0	0	0	0	0	0	0	0	0	0	-1	0	-1	0	0	0	0	0	1	0	0	0	0	0	0	0	0	0	0	0	0	-1	0	0
CO2	0	0	0	0	0	0	0	0	0	0	0	0	0	1	0	0	0	0	0	0	0	0	0	0	0	0	0	0	0	0	0	0	0	0	0	0	0	0	0	0	0	0	0	-1	0
NH3	0	0	0	0	0	0	0	0	0	0	0	0	0	0	0	0	0	0	0	0	0	1	1	0	0	0	0	0	0	0	0	0	0	0	0	0	0	0	0	0	0	0	0	0	-1
P1	0	0	0	0	0	0	0	0	0	0	0	0	0	0	0	0	0	0	0	0	0	0	0	0	0	0	0	0	0	2	0	0	0	0	0	0	0	0	0	0	0	-2	0	0	1
P2	0	0	0	0	0	0	0	0	0	0	0	0	0	0	0	0	0	0	0	0	0	0	0	0	0	0	0	0	0	0	0	6	6	0	0	0	0	0	0	0	0	13	3	3	0
P3	0	0	0	0	0	0	0	0	0	0	0	0	0	0	0	0	0	0	0	0	0	0	0	0	0	0	0	0	0	-1	0	2	2	0	0	0	0	0	0	0	0	4	3	0	0
P4	0	0	0	0	0	0	0	0	0	0	0	0	0	0	0	0	0	0	0	0	0	0	0	0	0	0	0	0	0	-3	0	2	2	0	0	0	0	0	0	0	0	4	0	0	0
P5	0	0	0	0	0	0	0	0	0	0	0	0	0	0	0	0	0	0	0	0	0	0	0	0	0	0	0	0	0	3	0	2	2	0	0	0	0	0	0	0	0	4	0	0	1
P6	0	0	0	0	0	0	0	0	0	0	0	0	0	0	0	0	0	0	0	0	0	0	0	0	0	0	0	0	0	3	0	0	0	0	0	0	0	0	0	0	0	0	0	0	1
P7	0	0	0	0	0	0	0	0	0	0	0	0	0	0	0	0	0	0	0	0	0	0	0	0	0	0	0	0	0	-2	0	0	0	0	0	0	0	0	0	0	0	0	0	0	0
P8	0	0	0	0	0	0	0	0	0	0	0	0	0	0	0	0	0	0	0	0	0	0	0	0	0	0	0	0	0	-1	0	0	0	0	0	0	0	0	0	0	0	0	0	0	0

rank of **S** is 37; thus, the left null space has a dimension of $40 - 37 = 3$ and the the right null space has a dimension of $45 - 37 = 8$. The left null space corresponds to the conservation of three cofactors: NADH, NADPH, and GR_{tot}.

Pathway vectors   There are eight pathway vectors, shown in Table 12.5, that are chosen to match those used above and in Chapter 11. p_1 is the glycolytic pathway to lactate secretion, p_2 is the pyruvate–lactate exchange coupled to the NADH cofactor, and p_3 is the pentose pathway cycle to CO_2 formation producing NADPH. These are the same pathways as above. In Chapter 11 we had AMP exchange that now couples to the five AMP degradation and salvage pathways that form in the AMP metabolic sub-network. They are as follows:

1. p_4 is a complicated pathway that imports glucose that is processed through the pentose pathway to form R5P and the formation of CO_2. R5P is then combined with an imported adenine to form AMP that then goes through the degradation pathway to adenosine that is secreted. This pathway is a balanced use of the whole network and represents a combination of the AMP biosynthetic and degradative pathways. This pathway corresponds to p_1 in the AMP sub-network.
2. p_5 is a similar pathway to p_4 above, but producing inosine, and it corresponds to p_2 in the AMP sub-network. Inosine is produced from AMP by first dephosphorylation to adenosine followed by deamination to inosine.
3. p_6 is a similar pathway to p_4 above, but producing inosine, and it corresponds to p_3 in the AMP sub-network. Inosine is produced from AMP by first deamination to IMP followed by dephosphorylation to inosine; the same effect as in p_5, but in the opposite order.
4. p_7 is the salvage pathway that has a net import of adenine and secreting hypoxanthine with a deamination step.
5. p_8 is a futile cycle that consumes ATP through the simultaneous use of AK and AMPase.

These pathway vectors that span the null space are shown pictorially in Figure 12.5. As we saw with the integration of glycolysis and the pentose pathway, the incorporation of the AMP sub-network creates new network-wide pathways. These network-based pathway definitions further show how historical definitions of pathways give way to full network considerations as we form pathways for the entire system. Again, this feature is of great importance to systems biology [89, 94].

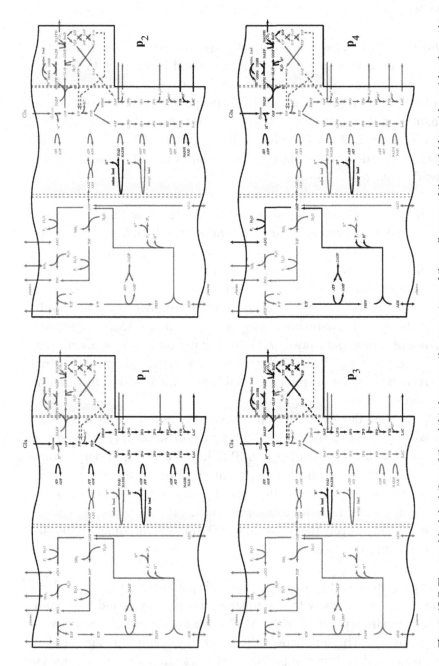

Figure 12.5 The graphical depiction of the eight chosen pathway vectors of the null space of the stoichiometric matrix, shown in Table 12.5, for the integrated model. The pathways are described in the text.

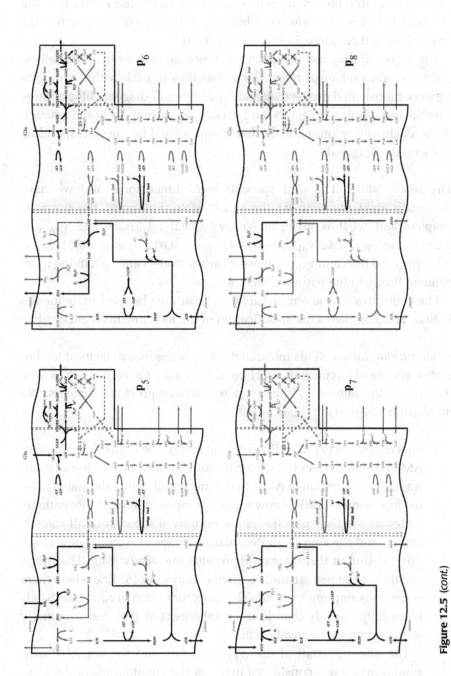

Figure 12.5 (cont.)

Structure of the stoichiometric matrix   The matrix can be organized by pathway and metabolic processes. The dashed lines in Table 12.5 divide the matrix up into blocks. Each block in this formulation shows how the metabolites and reactions are coupled. These diagonal blocks describe the metabolites and reactions in each sub-network.

The upper off-diagonal blocks, for instance, show how the metabolites, cofactors, and exchanges affect the metabolites in each pathway or sub-network represented on the diagonal. The lower off-diagonal blocks show how the metabolites, cofactors, and exchanges participate in the reaction in each pathway or sub-network that is represented by the corresponding block on the diagonal.

The steady state   The null space is eight-dimensional and we have to specify eight fluxes to fix the steady state. Following the previous chapters and what is given above, we specify: $v_{GLUin} = 1.12$, $v_{NADH} = 0.2v_{GLUin}$, $v_{GSHR} = 0.42$, $v_{ADE} = -0.014$, $v_{ADA} = 0.01$, $v_{ADO} = -0.01$, $v_{AK} = 0.12$, $v_{HYP} = 0.097$, and $v_{ApK} = 0$ using earlier models and specifying independent fluxes in the pathway vectors.

The concentrations given in Table 12.1 can now be used to define the PERCs. This completes the specification of the dynamic mass balances.

Dynamic simulation   This integrated network model can be used to simulate a variety of circumstances. Here, we simulate the response to a 50% increase in the rate of ATP usage as we have in previous chapters. We highlight two aspects of this simulation.

1. *Flux balancing of the AMP node:* Previously, the influx (formation) of AMP into the system has been constant, but the output (degradation) was a linear function of AMP. In the integrated model simulated here, the formation of AMP is now explicitly represented by a biosynthetic pathway. We thus plot the phase portrait of the sum of all the formation and the sum of all degradation fluxes of AMP; Figure 12.6a. The 45° line in this diagram represents the steady state. The initial reaction to the perturbation is motion above the 45° line where there is net consumption of AMP. The trajectory turns around and heads towards the steady-state line, overshoots it at first, but eventually settles down in the steady state.

   The phase portrait of the ATP load flux and the net AMP consumption rate was considered in the earlier chapters and is shown in Figure 12.6b. As before, the sudden increase in the ATP is followed by a net removal of AMP from the system and a dropping ATP load flux to reach the steady state again.

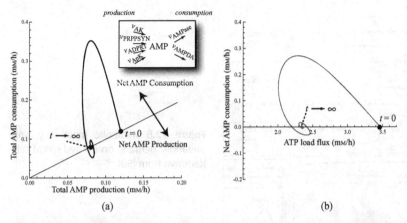

Figure 12.6 Dynamic response of the combined glycolytic, pentose pathway, and AMP metabolism network to a 50% increase in the rate of ATP use. (a) The phase portrait of total AMP consumption and production fluxes. The 45° line is the steady state, above which there is a net consumption (i.e., efflux) from the AMP node and below which there is a net production (i.e., import) of AMP. (b) The phase portrait of ATP load and net AMP consumption fluxes. See Figures 10.5a and 11.6a for comparison.

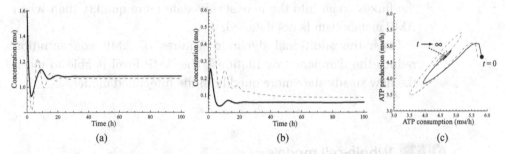

Figure 12.7 Dynamic response of the combined glycolytic and pentose pathway network (dashed line) and the glycolytic, pentose pathway, and AMP metabolism network (solid line) to a 50% increase in the rate of ATP use. (a) ATP concentration. (b) AMP concentration. (c) The dynamic phase portrait of ATP consumption and production fluxes.

2. *Comparison with the coupled glycolytic pentose pathway network from Chapter 11:* The phase portrait trajectory is qualitatively similar to simulated responses in previous chapters, where the nucleotide metabolism is not explicitly described. To get a quantitative comparison of the effects of detailing AMP metabolism, we simulate the glycolytic and pentose pathway model of Chapter 11 and compare it with the integrated model developed in this chapter (see Figure 12.7). The time response of ATP (Figure 12.7a) shows a more dampened response of the system with AMP metabolism versus the system without it. The same is true for the AMP response (Figure 12.7b) and the long-term transient is less pronounced. The phase portrait of the sum

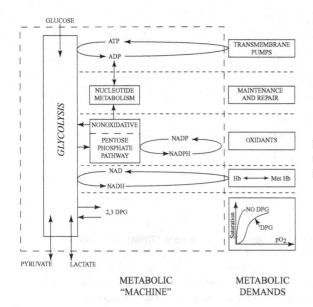

Figure 12.8 Metabolic machinery and metabolic demand on the red blood cell. Redrawn from [59].

of ATP consuming- and producing-fluxes (Figure 12.7c) shows how the fluxes come into the new steady state more quickly than when AMP metabolism is not detailed.

Thus, the additional dynamic features of AMP concentration reduce the dampened oscillations. The AMP level is able to reach the new steady state more quickly in the integrated model.

12.3 Whole-cell models

The network described herein encompasses the core metabolic functions in the human red blood cell. One can continue to build metabolic processes into this network model following the procedures outlined in this chapter. To build a more comprehensive model we would have to account for more details, such as osmotic pressure and electroneutrality [61]. The inclusion of these effects comes with complex mathematics, but accounts for very important physiological processes. The sodium–potassium pump would also have to be accounted for, as it is a key process in maintaining osmotic balance and cell shape. Another complication that arises pertains to the magnesium ion and the fact that it binds ATP to form MgATP (which is actually the substrate of many of the glycolytic enzymes) and to 23DPG to form Mg23DPG.

The metabolites also bind to macromolecules. Some macromolecules have many ligands, leading to a multiplicity of bound states. The bound states will alter the properties of the macromolecule. In Part IV we illustrate this phenomenon for hemoglobin and for a regulated enzyme, PFK.

These processes can be added in a stepwise fashion to increase the scope of the model and develop it towards a whole-cell model. Once comprehensive coverage of the known processes in a cell is achieved, a good physiological representation is obtained. For the simple red blood cell, such a representation is achievable and one can match the "metabolic machinery" with the "metabolic demands" that are placed on a cell (see Figure 12.8). Coordination is achieved through regulation. Clearly, the metabolic network is satisfying multiple functions simultaneously, an important feature of systems biology.

12.4 Summary

> Network reconstruction proceeds in a stepwise fashion by systematically integrating sub-networks. New issues may arise during each step of the integration process.

> The addition of the AMP sub-network introduces a few integration issues. These include the reactions that have common metabolites in the two networks being integrated, new plasma exchange reactions, and the detailing of stoichiometry that may have been simplified.

> The addition of the AMP synthesis and degradation sub-network introduces five new dimensions in the null space of **S**, where three represent the degradation of AMP and two represent the biosynthesis of AMP. New network-level pathways are introduced.

> One can continue to integrate more and more biochemical processes known to occur in a cell using the procedures outlined in this chapter. For simple cells, this process can approach a comprehensive description of the cell, and thus approach a whole-cell model. In addition to metabolic processes, a number of physico-chemical processes need to be added, and the functions of macromolecules can be incorporated, as described in the next part of this text.

> A large model can be analyzed and conceptualized at the three levels (biochemistry, systems biology, and physiology) as detailed for glycolysis in Chapter 10.

## PART IV

Macromolecules

In the fourth and final part of the book we develop MASS models that include macromolecular functions that are coupled with metabolites. Many proteins in cells bind to small molecules and the bound states of the proteins have different functions than the unbound forms.

The simplest and perhaps best known case is that of oxygen binding to hemoglobin. DPG23 is a unique metabolite in the red cell that binds to hemoglobin and alters the oxygen-carrying capacity of hemoglobin. DPG23 is made by a simple bypass on glycolysis, and in Chapter 13 we couple this bypass to the glycolytic model developed earlier in Chapter 10.

Furthermore, the kinases in glycolysis can bind to many metabolites. Some binding states inhibit enzyme catalysis and some activate function. In Chapter 14 we show how such regulatory models can be incorporated into the glycolytic model. With these examples, the reader should have the basic understanding of the procedure and data needs for building large-scale kinetic models in cell and molecular biology.

## CHAPTER 13

Hemoglobin

One of the primary functions of the red blood cell is to carry oxygen bound to hemoglobin. Hemoglobin can bind to a number of small molecules that affect its oxygen-carrying function. The binding of small molecules to a protein like hemoglobin can be described by chemical equations. All such chemical equations can be assembled into a stoichiometric matrix that describes all the bound states of the protein. Such a stoichiometric matrix can then be combined with another one that describes a metabolic network. In this chapter we describe the process that leads to an integrated kinetic description of low molecular weight metabolites and macromolecules, using hemoglobin as an example.

13.1 Hemoglobin: the carrier of oxygen

Physiological function and oxygen transport   Red blood cells account for approximately 45% of the volume of the blood, a number known as the *hematocrit*. With a blood volume of 5 L, a "standard man" [112] has about 2.25 L of red blood cells, or a total of approximately 2.5×10^{13} red cells. Each red blood cell contains approximately 30 pg of hemoglobin.

Oxygen is carried in the blood in two forms: (1) dissolved in the plasma; (2) bound to hemoglobin inside the red blood cell. Oxygen is a nonpolar molecule that dissolves poorly in aqueous solution. Its solubility is only about 7 ppm at 37 °C. Thus, 1 L of blood will only dissolve about 3.2 mL of oxygen. Conversely, the amount of oxygen carried by hemoglobin is much greater than the amount of oxygen dissolved in the plasma. The approximately 160 g/L of hemoglobin in blood will bind to the equivalent of about 220 mL of oxygen. Thus, the ratio between hemoglobin-bound oxygen to that dissolved in plasma is about 70:1. Hemoglobin thus enables oxygen to

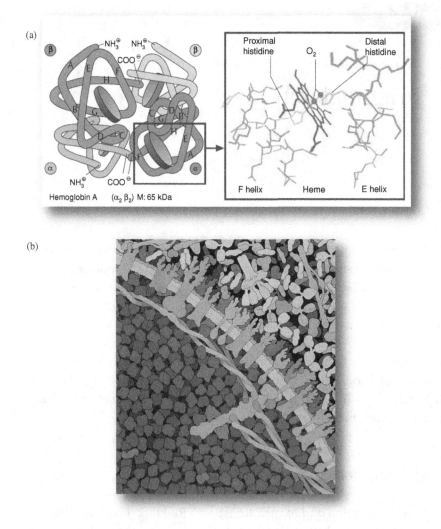

Figure 13.1 (a) The structure of human hemoglobin [68] (reprinted with permission). (b) An image of the packing density of hemoglobin in red cells ©David S. Goodsell 2000.

be transported to tissues at a high rate to meet their oxygen consumption rates. An interruption in oxygen delivery, known as an ischemic event, even for just a few minutes, can have serious consequences. Systemic and prolonged shortage of oxygen is know as anemia, and there are hundreds of millions of people on the planet that are anemic. Sufficient delivery of oxygen to tissues represents a critical physiological function.

Hemoglobin structure and function Hemoglobin is a relatively small globular protein that is a tetramer of two α globin and two β globin molecules. Its molecular weight is about 60 kDa (Figure 13.1a). It is

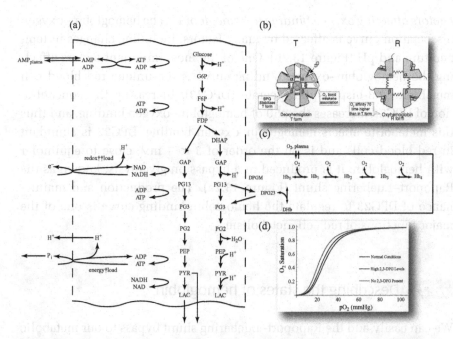

Figure 13.2 Glycolysis and the binding states of hemoglobin. (a) The formation of DPG23 through the Rapoport–Luebering bypass on glycolysis. (b) A schematic that shows how the binding of DPG23 displaces oxygen. From [68] (reprinted with permission). DPG23 is abbreviated as BPG in this panel. (c) The reaction schema for oxygen and DPG23 binding to hemoglobin. (d) Hemoglobin displays a sigmoidal, or S-shaped, oxygen-binding curve. Normal conditions: 3.1 mM; high DPG23: 6.0 mM, no DPG23 present: 0 mM.

by far the most abundant protein in the red blood cell, accounting for approximately 97% of its protein mass [17]. Its concentration is on the order of 7 mM in the red cell. Thus, red cells are packed with hemoglobin (Figure 13.1b).

Hemoglobin has four binding sites for oxygen. The oxygen binding site of hemoglobin is occupied by iron held in a heme group. The iron in hemoglobin represents 67% of the total iron in the human body. Hemoglobin–oxygen binding has a sigmoidal shape (Figure 13.2d) that results from the fact that hemoglobin's affinity for oxygen becomes greater as each binding site is filled with oxygen, a phenomenon known as *cooperativity*.

The binding curve is characterized by the p50, which is the partial pressure of oxygen at which hemoglobin is 50% saturated with oxygen. In other words, half of the oxygen binding sites are occupied at this partial pressure. The partial pressure of oxygen in venous blood is about 35–40 mmHg, whereas arterial blood in the lungs has a partial pressure of about 90–100 mmHg. Thus, hemoglobin leaves the lung saturated with oxygen and leaves the tissue at more than half saturation.

Factors affecting oxygen binding to hemoglobin The hemoglobin–oxygen dissociation curve is affected by many factors, including changes in temperature and pH (Figure 13.2b). One of the most important factors affecting the hemoglobin–oxygen binding curve is the unique red blood cell metabolite 2,3-bisphosphoglycerate (DPG23). Increasing the concentration of DPG23 increases the p50 of hemoglobin–oxygen binding, and thus this metabolite affects hemoglobin–oxygen binding. DPG23 is abundant in red blood cells and is on the order of 3 to 4 mM, close to equimolar with hemoglobin. It is produced in a bypass on glycolysis, known as the Rapoport–Luebering shunt (Figure 13.2a). The production and maintenance of DPG23 to regulate the hemoglobin-binding curve is one of the major functions of red cell metabolism.

13.2 Describing the states of hemoglobin

We can easily add the Rapoport–Luebering shunt bypass to our metabolic model, as shown in previous chapters for other metabolic processes. However, perhaps more importantly, we need to add hemoglobin as a compound to our network. We thus add a protein molecule to the network that we are considering, forming a set of reactions that involve both small metabolites and the proteins that they interact with. To do so, we need to consider the many states of hemoglobin.

The many states of hemoglobin Hemoglobin can bind to many ligands or small molecules. For illustrative purposes, we will consider two here: oxygen and DPG23. As shown in Figure 13.2b, the binding of DPG23 to hemoglobin prevents oxygen molecules from binding to hemoglobin. The corresponding reaction schema is shown in Figure 13.2c and it can be used to build a chemically detailed picture of this process. Here, we will assume that effectively all the oxygen is removed when DPG23 is bound.

The ligand binding reactions The binding of oxygen to hemoglobin proceeds serially, giving rise to four chemical reactions:

$$Hb_0 + O_2 \rightleftharpoons Hb_1, \tag{13.1}$$

$$Hb_1 + O_2 \rightleftharpoons Hb_2, \tag{13.2}$$

$$Hb_2 + O_2 \rightleftharpoons Hb_3, \tag{13.3}$$

$$Hb_3 + O_2 \rightleftharpoons Hb_4. \tag{13.4}$$

Table 13.1 The elemental matrix for the hemoglobin sub-network

	DPG23	O_2	Hb_0	Hb_1	Hb_2	Hb_3	Hb_4	DHb
C	3	0	0	0	0	0	0	3
H	3	0	0	0	0	0	0	3
O	10	2	0	2	4	6	8	10
P	2	0	0	0	0	0	0	2
Hb	0	0	1	1	1	1	1	1

As more oxygen molecules are bound, the easier a subsequent binding is, thus leading to cooperativity and a sigmoid binding curve; Figure 13.2d.

DPG23 can bind reversibly to Hb_0 to form the deoxy form DHb:

$$Hb_0 + DPG23 \rightleftharpoons DHb. \tag{13.5}$$

To form a network, we add the inputs and outputs for the ligands. DPG23 forms and degrades as

$$\xrightarrow{\text{DPGM}} DPG23 \xrightarrow{\text{DPGase}}. \tag{13.6}$$

DPG23 is formed by a mutase and degraded by a phosphatase. Interestingly, both these enzymatic activities are found on the same enzyme.

Oxygen enters and leaves the system by simple diffusion in and out of the red cell:

$$O_{2,\text{plasma}} \underset{\text{out}}{\overset{\text{in}}{\rightleftharpoons}} O_2. \tag{13.7}$$

The elemental matrix The elemental matrix for this system is in Table 13.1. Note that hemoglobin (Hb) is a chemical moiety treated as a whole. It is not formed or degraded in the reaction schema considered.

The stoichiometric matrix These reactions can be represented with a stoichiometric matrix; Table 13.2. The dimensions of this matrix are 8×8 and its rank is 7. There is one pathway, the Rapoport–Luebering shunt, that spans the null space of the stoichiometric matrix; see Table 13.2. There is one time-invariant, which is simply the total amount of hemoglobin presented by the last line in Table 13.1.

Pools and ratios Hemoglobin will have two basic forms: the oxy form (i.e., OHb bound to oxygen) and the deoxy form (i.e., DHb bound to

Table 13.2 The stoichiometric matrix, the elemental balancing of reactions, the pathways, and the steady-state flux vector for the hemoglobin binding system. The last column gives the computed steady state concentrations. The last row gives the equilibrium constants for the reactions in mmHg

	v_{DPGM}	v_{DPGase}	v_{HbO1}	v_{HbO2}	v_{HbO3}	v_{HbO4}	v_{HbDPG}	v_{O2}	Conc. (mM)
DPG23	1	−1	0	0	0	0	−1	0	3.10
O_2	0	0	−1	−1	−1	−1	0	−1	0.020
Hb_0	0	0	−1	0	0	0	−1	0	0.060
Hb_1	0	0	1	−1	0	0	0	0	0.050
Hb_2	0	0	0	1	−1	0	0	0	0.074
Hb_3	0	0	0	0	1	−1	0	0	0.26
Hb_4	0	0	0	0	0	1	0	0	6.80
DHb	0	0	0	0	0	0	1	0	0.046
C	3	−3	0	0	0	0	0	0	
H	3	−3	0	0	0	0	0	0	
O	10	−10	0	0	0	0	0	−2	
P	2	−2	0	0	0	0	0	0	
Hb	0	0	0	0	0	0	0	0	
p_1	1	1	0	0	0	0	0	0	
v_{stst}	0.441	0.441	0	0	0	0	0	0	
K_{eq}	0	0	41.84	73.21	178	1290	0.25	1.0	

DPG23). The total amount of oxygen carried on hemoglobin (i.e., the occupancy) at any given time is

$$OHb = Hb_1 + 2Hb_2 + 3Hb_3 + 4Hb_4, \tag{13.8}$$

while the oxygen carrying capacity is $4Hb_{tot}$. We can thus define the ratio

$$r_{OHb} = \frac{OHb}{4Hb_{tot}} = \frac{Hb_1 + 2Hb_2 + 3Hb_3 + 4Hb_4}{4Hb_{tot}} \tag{13.9}$$

as the fractional oxygen saturation or the oxygen charge of hemoglobin. The total amount of hemoglobin is a constant. Similarly, we can define the pool of DPG23 as

$$DPG23_{tot} = DHb + DPG23; \tag{13.10}$$

we can thus define the state of DPG23 as a regulator by the ratio

$$r_{DPG23} = \frac{DHb}{DPG23_{tot}} = \frac{DHb}{DHb + DPG23}. \tag{13.11}$$

These two ratios, r_{OHb} and r_{DPG23}, describe the two functional states of hemoglobin and they will respond to environmental and metabolic perturbations.

The steady state The binding of the two ligands, oxygen and DPG23, to hemoglobin is a rapid process. Since hemoglobin is confined to the red blood cell, we can use equilibrium assumptions for the binding reactions. The binding of oxygen is at equilibrium:

$$K_1 = \frac{Hb_1}{Hb_0 \times O_2} \quad K_2 = \frac{Hb_2}{Hb_1 \times O_2} \quad K_3 = \frac{Hb_3}{Hb_2 \times O_2} \quad K_4 = \frac{Hb_4}{Hb_3 \times O_2}.$$

The binding of DPG23 to hemoglobin is also at equilibrium:

$$K_d = \frac{DHb}{Hb_0 \times DPG23}. \tag{13.12}$$

The numerical values for the equilibrium constants are given in Table 13.2. The total mass of hemoglobin is a constant

$$Hb_{tot} = Hb_0 + Hb_1 + Hb_2 + Hb_3 + Hb_4 + DHb. \tag{13.13}$$

These six equations have six unknowns (the six forms of Hb) and need to be solved simultaneously as a function of the oxygen and DPG23 concentrations. The equilibrium relationships can be combined with the Hb_0 mass balance to

$$Hb_{tot} = Hb_0(1 + K_1O_2 + K_1K_2O_2^2 + K_1K_2K_3O_2^3 + K_1K_2K_3K_4O_2^4 + K_dDPG23).$$

This equation is solved for Hb_0 for given oxygen and DPG23 concentrations. Then, all the other forms of hemoglobin can be computed from the equilibrium relationships.

Numerical values The flux through the Rapoport–Luebering shunt is typically about 0.44 mM/h [110], leading to the steady-state flux vector shown in Table 13.2. The steady-state concentration of DPG23 is typically about 3.1 mM [77]. The equilibrium binding constants are also given in Table 13.2.

Using these numbers, the computed steady-state concentrations are obtained, as shown in Table 13.2, and the computed ratios are

$$r_{OHb} = 0.97, \quad r_{DPG23} = 0.015. \tag{13.14}$$

Thus, in this steady state, the oxygen-carrying capacity of hemoglobin is 97% utilized. Only 1.5% of the DPG23 is bound to hemoglobin. At this low DPG23 loading, the sequestration of hemoglobin into the deoxy form is maximally sensitive to changes in the concentration of the DPG23 regulator. Both of these features meet expectations of physiological function, i.e., almost full use of oxygen-carrying capacity while being maximally sensitive to a key regulatory signal.

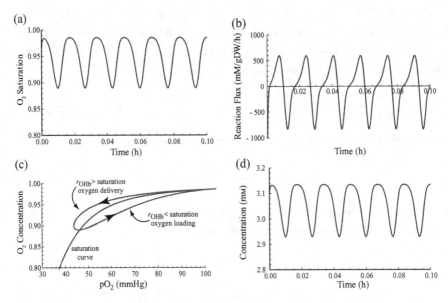

Figure 13.3 The binding states of hemoglobin during normal circulation. (a) r_{OHb}. (b) v_{O_2}. (c) r_{OHb} versus pO_2 in mmHg cycle shown with the hemoglobin saturation curve. (d) DPG23 concentration.

Dynamic responses The oxygenated state of hemoglobin will respond to the partial pressure of oxygen in the plasma. At sea level, the partial pressure of oxygen will change from a high of about 100 mmHg leaving the lung and 40 mmHg in venous blood leaving tissues. In the dynamic simulations below, we will use the average oxygen partial pressure of 70 mmHg and an oscillation from 40 to 100 mmHg as a time-variant environment during circulations. The average partial pressure can be dropped to simulate increased altitude.

The dynamic response of this system around the defined steady state can be performed. There are two cases of particular interest. First would be physiologically meaningful oscillations in the oxygen concentrations in the plasma; second is changes to the average oxygen concentration in the plasma that would, for instance, correspond to significant altitude changes. The former is rapid, while the latter is slow.

1. *Normal circulation* The average circulation time of a red cell in the human body is about 1 min. The partial pressure of oxygen in plasma should vary between 100 mmHg and 40 mmHg. To simulate the circulation cycle, we pick a sinusoidal oscillatory forcing function on the oxygen plasma and simulate the response. The simulated response of the state of the hemoglobin molecule is shown in Figure 13.3.

The fractional oxygen loading r_{OHb} shows how hemoglobin is highly saturated at 100 mmHg and that it drops to a low level at 40 mmHg;

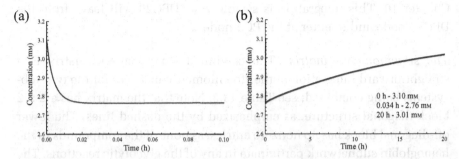

Figure 13.4 Dynamic response of the DPG23 concentration when the average partial pressure of oxygen drops from 70 mmHg at sea level to 38 mmHg at 12 000 ft. (a) Fast response. (b) Slow response.

Figure 13.3a. The corresponding exchange of oxygen in and out of the red cell shows how oxygen is flowing in and out of the red cell during the simulated circulation cycle; Figure 13.3b. Integrating the v_{O_2} curve shows that the oxygen loading and delivery is 2.8 mol/L per cycle. For a blood volume of 4.5 L, this corresponds to 12.6 mmol/cycle or 0.76 mol/h for 60 cycles per hour. This number is in the range of oxygen consumption for an adult male.

The loading and unloading of oxygen on hemoglobin can be visualized by graphing the oxygen loading r_{OHb} versus the plasma partial pressure of oxygen on the same plot as the saturation curve; Figure 13.3c. The displacement of DPG23 during circulation is shown in Figure 13.3d. We see how the loading and unloading of oxygen on hemoglobin is a cycle between the saturation curves that correspond to the maximum and minimum concentration of DPG23 during the cycle. Note how narrow the range of operation is, but that this is sufficient to get high rates of oxygen delivery from the lung to the tissues.

2. *Response to altitude change* The second situation of interest is a change in the average partial pressure of oxygen. Here, we drop it to 38 mmHg, representing a move to 12 000 ft elevation [4]. This change induces a long-term change in the DPG23 concentration (see Figure 13.4) that in turn induces a left shift in the oxygen–Hb binding curve; Figure 13.2.

13.3 Integration with glycolysis

The unique red blood cell metabolite DPG23 is produced by a bypass on glycolysis. This bypass can be integrated with the glycolytic model of

Chapter 10. This integration is seamless, as DPG23 will leave from the DPG13 node and re-enter at the PG3 node.

The stoichiometric matrix   The combined stoichiometric matrix is a straightforward integration of the stoichiometric matrices for the two sub-systems being combined; see Table 13.3. Note that the matrix has a 2×2 block diagonal structure, as emphasized by the dashed lines. The lower off-diagonal block has no nonzero entries, as none of the compounds in the hemoglobin subnetwork participate in any of the glycolytic reactions. The upper right off-diagonal block has the enzymes that make and degrade DPG23 and represent the coupling between the two sub-networks. The two diagonal blocks represent the two sub-networks and are the same as the stoichiometric matrices for the individual networks; see Tables 10.3 and 13.2.

The null space of the stoichiometric matrix   The null space of the integrated stoichiometric matrix is of dimension 4. The first three pathways are the same as for glycolysis alone (Table 13.3); recall Table 10.3. The fourth pathway represents flow through glycolysis and through the Rapoport–Luebering shunt, bypassing the ATP-generating PGK. This pathway thus produces no net ATP, but an inorganic phosphate with the degradation of DPG23. Thus, the DPG23 ligand that regulates the oxygen affinity of hemoglobin costs the red blood cell the opportunity to make one ATP.

Pools and ratios   The pools and ratios will be the same as for the two individual subsystems, with the exception of the role of DPG23. In the total phosphate pool, it will count as a 2, since it is two phosphate bonds. However, it has a high-energy bond value of 1 in the glycolytic energy pool (GP^+) as it can only generate one ATP through degradation through pyruvate kinase. In the glycolytic redox pool it has a value of zero, as it cannot generate NADH.

The steady state   The steady-state concentrations for the glycolytic inter-mediates can stay the same as before. The new steady-state concentration that needs to be added is that of DPG23. All glycolytic fluxes stay the same except for PGK. The PGK flux needs to be reduced by the amount that goes through the Rapoport–Luebering shunt: 0.441 mM/h. This creates minor changes in the estimated PERCs from this steady-state data: $k_{PGK} = 1.02 \times 10^6/(h\ mM)$ (compared with $1.274 \times 10^6/(h\ mM)$ in Table 10.5) and introduces $k_{DPGM} = 1814/h$, and $v_{DPGase} = 0.142/h$.

Dynamic simulation: normal circulation   We can repeat the response to the oscillatory plasma concentration of oxygen that we performed for

Table 13.3 The stoichiometric matrix for glycolysis, the Rapoport–Luebering (R–L) shunt, and the binding states of hemoglobin. The matrix is partitioned to show the states of hemoglobin separate from the glycolytic intermediates. The pathway vectors and the steady-state fluxes are show at the bottom of the table. p_4 is the new pathway vector

	Glycolysis																					R–L shunt		Hemoglobin					
	v_{HK}	v_{PGI}	v_{PFK}	v_{TPI}	v_{ALD}	v_{GAPDH}	v_{PGK}	v_{PGLM}	v_{ENO}	v_{PK}	v_{LDH}	v_{AMP}	v_{APK}	v_{PYR}	v_{LAC}	v_{ATP}	v_{NADH}	v_{GLUIN}	v_{AMPIN}	v_{H+}	v_{H2O}	v_{DPGM}	v_{DPGASE}	v_{HB01}	v_{HB02}	v_{HB03}	v_{HB04}	v_{HBDPG}	v_{O2}
GLU	−1	0	0	0	0	0	0	0	0	0	0	0	0	0	0	0	0	1	0	0	0	0	0	0	0	0	0	0	0
G6P	1	−1	0	0	0	0	0	0	0	0	0	0	0	0	0	0	0	0	0	0	0	0	0	0	0	0	0	0	0
F6P	0	1	−1	0	0	0	0	0	0	0	0	0	0	0	0	0	0	0	0	0	0	0	0	0	0	0	0	0	0
FBP	0	0	1	0	−1	0	0	0	0	0	0	0	0	0	0	0	0	0	0	0	0	0	0	0	0	0	0	0	0
DHAP	0	0	0	−1	1	0	0	0	0	0	0	0	0	0	0	0	0	0	0	0	0	0	0	0	0	0	0	0	0
GAP	0	0	0	1	1	−1	0	0	0	0	0	0	0	0	0	0	0	0	0	0	0	0	0	0	0	0	0	0	0
PG13	0	0	0	0	0	1	−1	0	0	0	0	0	0	0	0	0	0	0	0	0	0	−1	0	0	0	0	0	0	0
PG3	0	0	0	0	0	0	1	−1	0	0	0	0	0	0	0	0	0	0	0	0	0	0	1	0	0	0	0	0	0
PG2	0	0	0	0	0	0	0	1	−1	0	0	0	0	0	0	0	0	0	0	0	0	0	0	0	0	0	0	0	0
PEP	0	0	0	0	0	0	0	0	1	−1	0	0	0	0	0	0	0	0	0	0	0	0	0	0	0	0	0	0	0
PYR	0	0	0	0	0	0	0	0	0	1	−1	0	0	−1	0	0	0	0	0	0	0	0	0	0	0	0	0	0	0
LAC	0	0	0	0	0	0	0	0	0	0	1	0	0	0	−1	0	0	0	0	0	0	0	0	0	0	0	0	0	0
NAD	0	0	0	0	0	−1	0	0	0	0	1	0	0	0	0	0	1	0	0	0	0	0	0	0	0	0	0	0	0
NADH	0	0	0	0	0	1	0	0	0	0	−1	0	0	0	0	0	−1	0	0	0	0	0	0	0	0	0	0	0	0
AMP	0	0	0	0	0	0	0	0	0	0	0	−1	1	0	0	0	0	0	1	0	0	0	0	0	0	0	0	0	0
ADP	1	0	1	0	0	0	−1	0	0	−1	0	0	−2	0	0	1	0	0	0	0	0	0	0	0	0	0	0	0	0

(continued)

Table 13.3 (continued).

| | Glycolysis | | | | | | | | | | | | | | | | | | | | | R–L shunt | | Hemoglobin | | | | | |
	v_{HK}	v_{PGI}	v_{PFK}	v_{TPI}	v_{ALD}	v_{GAPDH}	v_{PGK}	v_{PGLM}	v_{ENO}	v_{PK}	v_{LDH}	v_{AMP}	v_{ApK}	v_{PYR}	v_{LAC}	v_{ATP}	v_{NADH}	v_{GLUIN}	v_{AMPIN}	v_{H^+}	v_{H_2O}	v_{DPGM}	v_{DPGASE}	v_{HB01}	v_{HB02}	v_{HB03}	v_{HB04}	v_{HBDPG}	v_{O2}
ATP	−1	0	−1	0	0	0	1	0	0	1	0	0	1	0	0	−1	0	0	0	0	0	0	0	0	0	0	0	0	0
P_i	0	0	0	0	0	−1	0	0	0	0	0	0	0	0	0	1	0	0	0	0	0	0	1	0	0	0	0	0	0
H	1	0	1	0	0	1	0	0	0	−1	−1	0	0	0	0	−1	1	0	0	−1	0	−1	0	0	0	0	0	0	0
H_2O	0	0	0	0	0	0	0	0	1	0	0	0	0	0	0	0	0	0	0	0	−1	0	0	0	0	0	0	0	0
DPG23	0	0	0	0	0	0	0	0	0	0	0	0	0	0	0	0	0	0	0	0	0	1	−1	0	0	0	0	−1	0
O_2	0	0	0	0	0	0	0	0	0	0	0	0	0	0	0	0	0	0	0	0	0	0	0	−1	−1	−1	−1	0	−1
Hb_0	0	0	0	0	0	0	0	0	0	0	0	0	0	0	0	0	0	0	0	0	0	0	0	−1	0	0	0	−1	0
Hb_1	0	0	0	0	0	0	0	0	0	0	0	0	0	0	0	0	0	0	0	0	0	0	0	1	−1	0	0	0	0
Hb_2	0	0	0	0	0	0	0	0	0	0	0	0	0	0	0	0	0	0	0	0	0	0	0	0	1	−1	0	0	0
Hb_3	0	0	0	0	0	0	0	0	0	0	0	0	0	0	0	0	0	0	0	0	0	0	0	0	0	1	−1	0	0
Hb_4	0	0	0	0	0	0	0	0	0	0	0	0	0	0	0	0	0	0	0	0	0	0	0	0	0	0	1	0	0
DHb	0	0	0	0	0	0	0	0	0	0	0	0	0	0	0	0	0	0	0	0	0	0	0	0	0	0	0	1	0
$\mathbf{P_1}$	1	1	1	1	1	2	2	2	2	2	2	0	0	0	2	2	0	1	0	2	0	0	0	0	0	0	0	0	0
$\mathbf{P_2}$	0	0	0	0	0	0	0	0	0	0	−1	0	0	1	−1	0	−1	1	0	2	0	0	0	0	0	0	0	0	0
$\mathbf{P_3}$	0	0	0	0	0	0	0	0	0	0	0	1	0	0	0	0	0	0	1	0	0	0	0	0	0	0	0	0	0
$\mathbf{P_4}$	1	1	0	1	1	2	2	2	2	2	2	0	0	0	2	0	0	1	0	2	0	2	2	0	0	0	0	0	2
$\mathbf{v_{sst}}$	1.12	1.12	1.12	1.12	1.12	2.24	1.77	2.24	2.24	2.24	2.016	0.014	0	0.224	2.016	1.77	0.224	1.12	0.014	2.69	0	0.44	0.44	0	0	0	0	0	0

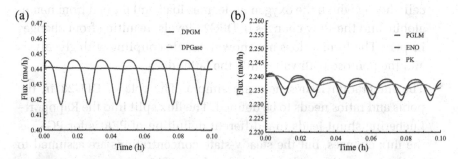

Figure 13.5 The binding states of hemoglobin during normal circulation as computed from the integrated glycolytic and Rapoport–Luebering shunt model. (a) The fluxes through the Rapoport–Luebering shunt. (b) The lower glycolytic fluxes.

the hemoglobin subsystem alone (Figure 13.3). The response of the integrated system demonstrates dynamic decoupling between the loading and offloading of oxygen on hemoglobin and glycolytic activity.

The simulation demonstrates dynamic decoupling at the rapid time scales, namely that the loading and offloading of oxygen on the minute time scale does not have significant ripple effects into glycolytic functions. The oxygen delivery and hemoglobin are unaltered, as compared before (Figure 13.3a and b). The input into the Rapoport–Luebering shunt is effectively a constant, while the output is slightly oscillatory (Figure 13.5a), leading to minor downstream effects on lower glycolysis (Figure 13.5b). Notice that the PK flux is barely perturbed.

13.4 Summary

> The ligand binding to macromolecules can be described by chemical equations. Stoichiometric matrices that describe the binding states of macromolecules can be formulated. Such matrices can be integrated with stoichiometric matrices describing metabolic networks, thus forming integrated models of small molecules and protein.

> Since hemoglobin is confined to the system, the rapid binding of the ligands lead to equilibrium states. These are not quasi-equilibria, since the serial binding is a "dead end," meaning there is no output on the other end of the series of binding steps. Thus, the ligand bound states of the macromolecule are equilibrium variables that match the steady-state values of the ligands. The ligands are in pathways that leave and enter the system and are thus in a steady state.

> The oxygen loading of hemoglobin has fast and slow physiologically meaningful time scales: the 1 min circulation time of the red blood

cell, during which the oxygen molecules load and unload from hemo-globin, and the slow changes in DPG23 levels, resulting from changes in pO_2. The former does not show dynamic coupling with glycolysis and the pentose pathway, while the latter does.

➤ The integration issues here are simple. The role of DPG23 in the pools and ratios needs to be defined. The flux split into the Rapoport–Luebering shunt leads to a different definition of PERCs for PGK, as the flux changes, but the steady-state concentrations are assumed to be the same with and without the shunt. This easy integration is a reflection of the MASS procedure and its flexibility.

## CHAPTER 14

Regulated enzymes

The proteins that are of particular interest from a dynamic simulation perspective are those that regulate fluxes. Regulatory enzymes bind to many ligands. Their functional states can be described in a similar fashion as done for hemoglobin. Their states and functions can thus be readily integrated into MASS models. In this chapter we detail the molecular mechanisms for a key regulatory enzyme in glycolysis, PFK. First, we describe the enzyme, then we detail the sub-network that it represents, and finally we integrate them into the metabolic model. We then simulate the altered dynamic network states that result from these regulatory interactions.

14.1 Phosphofructokinase

The enzyme   PFK is a tetrameric enzyme. There are isoforms of the subunits of the enzyme, meaning that there is more than one gene for each subunit and the genes are not identical. The isoforms are differentially expressed in various tissues; therefore, different versions of the enzyme are active in different tissues. The regulation of PFK is quite complicated [87]. Here, for illustrative purposes, we will consider a homotetrameric form of PFK, Figure 14.1, with one activator (AMP) and one inhibitor (ATP).

The reaction catalyzed   PFK is a major regulatory enzyme in glycolysis. It catalyzes the reaction

$$F6P + ATP \xrightarrow{PFK} FDP + ADP + H. \tag{14.1}$$

In Chapter 10 we introduced this reaction as a part of the glycolytic pathway.

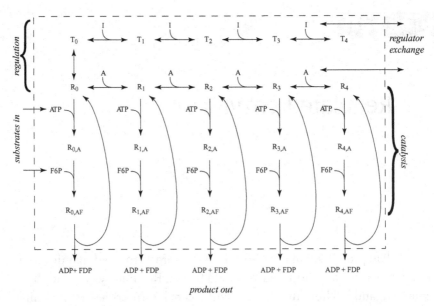

Figure 14.1 The PFK subnetwork; the reaction schema for catalysis, for regulation, and the exchanges with the rest of the metabolic network. The numerical values of the dissociation constants are taken from [60]; K_{ATP} is 0.068 mM, K_{F6P} is 0.1 mM, K_a is 0.33 mM, and K_i is 0.01 mM. The K_i binding constant for ATP as an inhibitor is increased by a factor of 10, since magnesium complexing of ATP is not considered here. The allosteric constant L is 0.0011.

The subnetwork The detailed reaction mechanism of PFK is shown in Figure 14.1. The reactants, F6P and ATP, and products, FDP and ADP, enter and leave the sub-network. These exchanges will reach a steady state. The regulators, AMP is an activator and ATP is an inhibitor, also enter and leave the sub-network. They reach an equilibrium in the steady state, as they have no flow through the sub-network. The bound states of PFK thus equilibrate with its regulators, while the reactants and products will flow through the subnetwork.

Bound states of PFK The enzyme exists in two natural forms: T, or tight, and R, or relaxed (recall Section 5.5):

$$R_0 \rightleftharpoons T_0. \tag{14.2}$$

The R form is catalytically active and the T form is inactive. The two forms are in equilibrium as

$$L = \frac{T_0}{R_0} = 0.0011, \tag{14.3}$$

where L is the allosteric constant for PFK.

There are many ligands that can bind to PFK to modulate its activity by altering the balance of the T and R forms. Here, we will consider AMP as an activator and ATP as an inhibitor. AMP will bind to the R state:

$$R_0 + AMP \rightleftharpoons R_1, \tag{14.4}$$

$$R_1 + AMP \rightleftharpoons R_2, \tag{14.5}$$

$$R_2 + AMP \rightleftharpoons R_3, \tag{14.6}$$

$$R_3 + AMP \rightleftharpoons R_4; \tag{14.7}$$

and ATP will bind to the T state:

$$T_0 + ATP \rightleftharpoons T_1, \tag{14.8}$$

$$T_1 + ATP \rightleftharpoons T_2, \tag{14.9}$$

$$T_2 + ATP \rightleftharpoons T_3, \tag{14.10}$$

$$T_3 + ATP \rightleftharpoons T_4. \tag{14.11}$$

The chemical reaction, see Eq. (14.1), will proceed in three steps:

1. the binding of the cofactor ATP

$$ATP + R_i \xrightarrow{\text{substrate in}} \rightleftharpoons R_{i,A}; \tag{14.12}$$

2. the binding of the substrate F6P

$$F6P + R_{i,A} \xrightarrow{\text{substrate in}} \rightleftharpoons R_{i,AF}; \tag{14.13}$$

3. the catalytic conversion to the products

$$R_{i,AF} \xrightarrow{\text{transformation}} R_i + FDP + ADP + H \xrightarrow{\text{products out}} \tag{14.14}$$

   and the release of the enzyme.

The subscript i ($i = 0, 1, 2, 3, 4$) indicates how many activator molecules (AMP) are bound to the R form of the enzyme. The free form ($i = 0$) and all the bound forms are catalytically active.

The elemental matrix   The elemental composition of the compounds in the PFK sub-network are shown in Table 14.1. The low molecular weight compounds are detailed in terms of their elemental composition, while the enzyme itself is treated as one moiety. It never leaves the system and is never chemically modified, so we can treat it as one entity.

The stoichiometric matrix   The reactions that correspond to this sub-network of binding states and catalytic conversions can be summarized in a stoichiometric matrix, shown in Table 14.2. This matrix is 26 × 30 of rank 25.

Table 14.1 The elemental composition of the compounds in the PFK subnetwork. The PFK molecule is treated as one group. The compounds are separated into the metabolites, the relaxed forms of the enzyme, and the tight forms of the enzyme

	Metabolites						R-form															T-form				
	$F6P$	FDP	AMP	ADP	ATP	H	R_0	$R_{0,A}$	$R_{0,AF}$	R_1	$R_{1,A}$	$R_{1,AF}$	R_2	$R_{2,A}$	$R_{2,AF}$	R_3	$R_{3,A}$	$R_{3,AF}$	R_4	$R_{4,A}$	$R_{4,AF}$	T_0	T_1	T_2	T_3	T_4
C	6	6	10	10	10	0	0	10	16	10	20	26	20	30	36	30	40	46	40	50	56	0	10	20	30	40
H	11	10	13	13	13	1	0	13	24	13	26	37	26	39	50	39	52	63	52	65	76	0	13	26	39	52
O	9	12	7	10	13	0	0	13	22	7	20	29	14	27	36	21	34	43	28	41	50	0	13	26	39	52
P	1	2	1	2	3	0	0	3	4	1	4	5	2	5	6	3	6	7	4	7	8	0	3	6	9	12
N	0	0	5	5	5	0	0	5	5	5	10	10	10	15	15	15	20	20	20	25	25	0	5	10	15	20
PFK	0	0	0	0	0	0	1	1	1	1	1	1	1	1	1	1	1	1	1	1	1	1	1	1	1	1

The null space is five-dimensional ($= 30 - 25$). The five pathway vectors spanning the null space are simply the input of the reagents, the PFK reaction itself, and the output of the products; see Table 14.2. Each pathway is catalyzed by one of the five active R forms. Thus, these form five parallel pathways through the sub-network, where the flux through each is determined by the relative amount of the R forms of the enzyme.

The left null space is one-dimensional ($= 26 - 25$). It is spanned by a vector that represents the conservation of the PFK enzyme itself.

Pools and ratios   The enzyme is in two key states, the relaxed (R) and the tight (T) state. The binding of the activator, AMP, pulls the enzyme towards the catalytically active relaxed state, while binding of the inhibitor, ATP, pulls it towards the catalytically inactive tight state, creating a complex tug of war over the catalytic capacity of the enzyme. This is presented pictorially in Figure 14.2. The fraction of the total enzyme that is in the R form can be computed:

$$r_R = \frac{\sum_{i=0}^{4}(R_i + R_{i,A} + R_{i,AF})}{PFK_{tot}}. \tag{14.15}$$

The relaxed states are then pooled into the form that is loaded with substrates that carry out the reaction. We thus define the fraction of the enzyme R states that is loaded with the substrates and is producing the product as

$$r_{cat} = \frac{\sum_{i=0}^{4} R_{i,AF}}{\sum_{i=0}^{4}(R_i + R_{i,A} + R_{i,AF})}. \tag{14.16}$$

This fraction of the enzyme generates the products.

Table 14.2 The stoichiometric matrix for the PFK subnetwork. The first six columns (up to the first dashed vertical line) are the input and output rates. The next set of columns are the PFK reaction itself (five sets, one for each R_i, $i = 0, 1, 2, 3, 4$). The last five columns are conversion into the T form and the binding reactions for the inhibitor. The first six rows are the compounds participating in the reaction, above the first horizontal dashed line, then the bound states of the enzyme, above the second horizontal dashed line, then the elemental balancing of the reactions, and finally the pathway vectors

	Input/output						Activation and catalysis																			Inhibition				
	v_{F6P}	v_{FDP}	v_{AMP}	v_{ADP}	v_{ATP}	v_{H}	$v_{R_{0,1}}$	$v_{R_{0,2}}$	$v_{R_{0,3}}$	$v_{R_{1,0}}$	$v_{R_{1,1}}$	$v_{R_{1,2}}$	$v_{R_{1,3}}$	$v_{R_{2,0}}$	$v_{R_{2,1}}$	$v_{R_{2,2}}$	$v_{R_{2,3}}$	$v_{R_{3,0}}$	$v_{R_{3,1}}$	$v_{R_{3,2}}$	$v_{R_{3,3}}$	$v_{R_{4,0}}$	$v_{R_{4,1}}$	$v_{R_{4,2}}$	$v_{R_{4,3}}$	v_{L}	v_{T_1}	v_{T_2}	v_{T_3}	v_{T_4}
F6P	1	0	0	0	0	0	0	-1	0	0	0	-1	0	0	0	-1	0	0	0	-1	0	0	0	-1	0	0	0	0	0	0
FDP	0	-1	0	0	0	0	0	0	1	0	0	0	1	0	0	0	1	0	0	0	1	0	0	0	1	0	0	0	0	0
AMP	0	0	1	0	0	0	-1	0	0	0	-1	0	0	0	-1	0	0	0	-1	0	0	0	-1	0	0	0	0	0	0	0
ADP	0	0	0	-1	0	0	0	0	1	0	0	0	1	0	0	0	1	0	0	0	1	0	0	0	1	0	0	0	0	0
ATP	0	0	0	0	1	0	0	0	-1	0	0	0	-1	0	0	0	-1	0	0	0	-1	0	0	0	-1	0	-1	-1	-1	-1
H	0	0	0	0	0	-1	0	0	1	0	0	0	1	0	0	0	1	0	0	0	1	0	0	0	1	0	0	0	0	0
R_0	0	0	0	0	0	0	-1	0	1	-1	0	0	0	0	0	0	0	0	0	0	0	0	0	0	0	-1	0	0	0	0
$R_{0,A}$	0	0	0	0	0	0	1	-1	0	0	0	0	0	0	0	0	0	0	0	0	0	0	0	0	0	0	0	0	0	0
$R_{0,AF}$	0	0	0	0	0	0	0	1	-1	0	0	0	0	0	0	0	0	0	0	0	0	0	0	0	0	0	0	0	0	0
R_1	0	0	0	0	0	0	0	0	0	1	-1	0	1	-1	0	0	0	0	0	0	0	0	0	0	0	0	0	0	0	0
$R_{1,A}$	0	0	0	0	0	0	0	0	0	0	1	-1	0	0	0	0	0	0	0	0	0	0	0	0	0	0	0	0	0	0
$R_{1,AF}$	0	0	0	0	0	0	0	0	0	0	0	1	-1	0	0	0	0	0	0	0	0	0	0	0	0	0	0	0	0	0
R_2	0	0	0	0	0	0	0	0	0	0	0	0	0	1	-1	0	1	-1	0	0	0	0	0	0	0	0	0	0	0	0
$R_{2,A}$	0	0	0	0	0	0	0	0	0	0	0	0	0	0	1	-1	0	0	0	0	0	0	0	0	0	0	0	0	0	0
$R_{2,AF}$	0	0	0	0	0	0	0	0	0	0	0	0	0	0	0	1	-1	0	0	0	0	0	0	0	0	0	0	0	0	0
R_3	0	0	0	0	0	0	0	0	0	0	0	0	0	0	0	0	0	1	-1	0	1	-1	0	0	0	0	0	0	0	0
$R_{3,A}$	0	0	0	0	0	0	0	0	0	0	0	0	0	0	0	0	0	0	1	-1	0	0	0	0	0	0	0	0	0	0
$R_{3,AF}$	0	0	0	0	0	0	0	0	0	0	0	0	0	0	0	0	0	0	0	1	-1	0	0	0	0	0	0	0	0	0
R_4	0	0	0	0	0	0	0	0	0	0	0	0	0	0	0	0	0	0	0	0	0	1	-1	0	1	0	0	0	0	0

(continued)

Table 14.2 (continued).

	Input/output						Activation and catalysis																			Inhibition				
	v_{F6P}	v_{FDP}	v_{AMP}	v_{ADP}	v_{ATP}	v_H	$v_{R0.1}$	$v_{R0.2}$	$v_{R0.3}$	$v_{R1.0}$	$v_{R1.1}$	$v_{R1.2}$	$v_{R1.3}$	$v_{R2.0}$	$v_{R2.1}$	$v_{R2.2}$	$v_{R2.3}$	$v_{R3.0}$	$v_{R3.1}$	$v_{R3.2}$	$v_{R3.3}$	$v_{R4.0}$	$v_{R4.1}$	$v_{R4.2}$	$v_{R4.3}$	v_L	v_{T1}	v_{T2}	v_{T3}	v_{T4}
$R_{4,A}$	0	0	0	0	0	0	0	0	0	0	0	0	0	0	0	0	0	0	0	0	0	0	1	−1	0	0	0	0	0	0
$R_{4,AF}$	0	0	0	0	0	0	0	0	0	0	0	0	0	0	0	0	0	0	0	0	0	0	0	1	−1	0	0	0	0	0
T_0	0	0	0	0	0	0	0	0	0	0	0	0	0	0	0	0	0	0	0	0	0	0	0	0	0	1	−1	0	0	0
T_1	0	0	0	0	0	0	0	0	0	0	0	0	0	0	0	0	0	0	0	0	0	0	0	0	0	0	1	−1	0	0
T_2	0	0	0	0	0	0	0	0	0	0	0	0	0	0	0	0	0	0	0	0	0	0	0	0	0	0	0	1	−1	0
T_3	0	0	0	0	0	0	0	0	0	0	0	0	0	0	0	0	0	0	0	0	0	0	0	0	0	0	0	0	1	−1
T_4	0	0	0	0	0	0	0	0	0	0	0	0	0	0	0	0	0	0	0	0	0	0	0	0	0	0	0	0	0	1
C	6	−6	10	−10	10	0	0	0	0	0	0	0	0	0	0	0	0	0	0	0	0	0	0	0	0	0	0	0	0	0
H	11	−10	13	−13	13	−1	0	0	0	0	0	0	0	0	0	0	0	0	0	0	0	0	0	0	0	0	0	0	0	0
O	9	−12	7	−10	13	0	0	0	0	0	0	0	0	0	0	0	0	0	0	0	0	0	0	0	0	0	0	0	0	0
P	1	−2	1	−2	3	0	0	0	0	0	0	0	0	0	0	0	0	0	0	0	0	0	0	0	0	0	0	0	0	0
N	0	0	5	−5	5	0	0	0	0	0	0	0	0	0	0	0	0	0	0	0	0	0	0	0	0	0	0	0	0	0
PFK	−1	1	0	0	0	1	−1	−1	−1	0	0	0	0	0	0	0	0	0	0	0	0	0	0	0	0	0	0	0	0	0
P_1	−1	1	0	1	−1	1	0	0	0	1	1	1	1	0	0	0	0	0	0	0	0	0	0	0	0	0	0	0	0	0
P_2	−1	1	0	1	−1	1	0	0	0	0	0	0	0	1	1	1	1	0	0	0	0	0	0	0	0	0	0	0	0	0
P_3	−1	1	0	1	−1	1	0	0	0	0	0	0	0	0	0	0	0	1	1	1	1	0	0	0	0	0	0	0	0	0
P_4	−1	1	0	1	−1	1	0	0	0	0	0	0	0	0	0	0	0	0	0	0	0	1	0	0	0	0	0	0	0	0
P_5	−1	1	0	1	−1	1	0	0	0	0	0	0	0	0	0	0	0	0	0	0	0	0	1	1	1	0	0	0	0	0

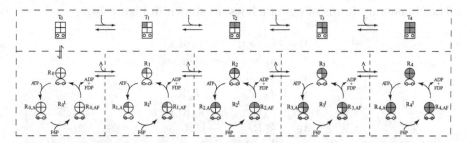

Figure 14.2 Pictorial representation of the PFK subnetwork. The inactive taut (T_l) states are designated with a square, and the substrate binding sites, designated with an open circle, are not accessible. The shading of the sub-squares indicates that the inhibitor (ATP) occupies that site. The relaxed forms (R_l) are shown with a circle, where the binding sites for the activator (AMP) are indicated. Shading means that the AMP binding site is occupied. The semi-circle below the circle shows that the binding sites are now on the surface of the protein and can now be occupied by the reactants (ATP and F6P). The $R_{l,AF}$ state catalyzes the reaction.

14.2 The steady state

The reaction rate The steady-state flux through the PFK subnetwork is given by

$$v_{\text{PFK}} = k_{\text{PFK}} \sum_{i=0}^{4} R_{i,\text{AF}} \tag{14.17}$$

$$= k_{\text{PFK}} \text{PFK}_{\text{tot}} r_{\text{R}} r_{\text{cat}}. \tag{14.18}$$

In this equation we know the steady-state flux and the total amount of enzyme. We will use the glycolytic flux in Chapter 10 of 1.12 mM/h. The concentration of PFK in the red blood cell is about 33 nM [6].

Integration of the PFK subnetwork into the metabolic network We can compute r_{R} and r_{cat} from equilibrium binding of the regulatory ligands to PFK. We know the steady-state concentrations of these ligands in the network; ATP is about 1.6 mM and AMP is about 0.087 mM. We also know the steady-state concentration of the reactants ATP and F6P (0.02 mM). The algebra associated with these computations is quite intricate, albeit simple in principle. Its complexity was foreshadowed in Chapter 5.

Once we have computed the PERC, k_{PFK}, we can integrate the PFK subnetwork into the glycolytic model. Such integration procedures can be performed for any enzyme of interest in a given network.

Solving for the steady state The steady-state equations are

$$\mathbf{S}\mathbf{v}(\mathbf{x}) = 0. \tag{14.19}$$

The elementary forms of the rate laws, as shown in Figure 14.2,

$$
\begin{aligned}
v_{R_{0,1}} &: k_A^+ R_0 & &- k_A^- R_{0,A}, \\
v_{R_{0,2}} &: k_F^+ R_{0,A} & &- k_F^- R_{0,AF}, \\
v_{R_{0,3}} &: k_{PFK} R_{0,AF} \\
v_{R_{1,0}} &: 4k_a^+ R_0 & &- k_a^- R_1, \\
v_{R_{1,1}} &: k_A^+ R_1 & &- k_A^- R_{1,A}, \\
v_{R_{1,2}} &: k_F^+ R_{1,A} & &- k_F^- R_{1,AF}, \\
v_{R_{1,3}} &: k_{PFK} R_{1,AF}, \\
v_{R_{2,0}} &: 3k_a^+ R_1 & &- 2k_a^- R_2, \\
v_{R_{2,1}} &: k_A^+ R_2 & &- k_A^- R_{2,A}, \\
v_{R_{2,2}} &: k_F^+ R_{2,A} & &- k_F^- R_{2,AF}, \\
v_{R_{2,3}} &: k_{PFK} R_{2,AF}, \\
v_{R_{3,0}} &: 2k_a^+ R_2 & &- 3k_a^- R_3, \\
v_{R_{3,1}} &: k_A^+ R_3 & &- k_A^- R_{3,A}, \\
v_{R_{3,2}} &: k_F^+ R_{3,A} & &- k_F^- R_{3,AF}, \\
v_{R_{3,3}} &: k_{PFK} R_{3,AF}, \\
v_{R_{4,0}} &: k_a^+ R_3 & &- 4k_a^- R_4, \\
v_{R_{4,1}} &: k_A^+ R_4 & &- k_A^- R_{4,A}, \\
v_{R_{4,2}} &: k_F^+ R_{4,A} & &- k_F^- R_{4,AF}, \\
v_{R_{4,3}} &: k_{PFK} R_{4,AF}, \\
v_L &: k^+ R_0 & &- k^- T_0, \\
v_{T_1} &: 4k_i^+ T_0 & &- k_i^- T_1, \\
v_{T_2} &: 3k_i^+ T_1 & &- 2k_i^- T_2, \\
v_{T_3} &: 2k_i^+ T_2 & &- 3k_i^- T_3, \\
v_{T_4} &: k_i^+ T_3 & &- 4k_i^- T_4
\end{aligned}
\tag{14.20}
$$

can be introduced into this equation to form 19 algebraic equations in 20 concentration variables as unknowns, where

$$
L = \frac{k^+}{k^-}, \quad K_i = \frac{k_i^+}{k_i^-}, \quad K_a = \frac{k_a^+}{k_a^-}, \quad K_F = \frac{k_F^+}{k_F^-}, \quad K_A = \frac{k_A^+}{k_A^-}
\tag{14.21}
$$

and where the forward rate constants k_F^+, k_A^+, k_a^+, and k_i^+ contain the steady-state concentrations of the corresponding ligand. This makes all the rate laws linear. k_A is the rate constant for the binding of ATP to the relaxed/active enzyme (R) during the catalytic process, k_F is the rate constant for the binding of F6P to the relaxed/active enzyme which already has ATP bound (R_A) during the catalytic process, k_a is the rate constant for the binding of AMP to the relaxed/active enzyme for activation of the enzyme, k_i is the rate constant for the binding of ATP to the taut/inhibited enzyme for inhibition of the enzyme, k_{PFK} is the rate constant for the release of the products (ADP and FDF) from R_{AF} during the catalytic process, and k_L is the rate constant for the interconversion of the relaxed/active and taut/inhibited enzyme.

Table 14.3 The steady-state solution for the enzyme concentrations in the PFK sub-network

i	$R_{i,0}$	$R_{i,0}/R_{i,0}$	$R_{i,A}/R_{i,0}$	$R_{i,AF}/R_{i,0}$	$R_{i,0}/R_{1,0}$	T_i/T_0
0	$\dfrac{(k_a^-)^4 v_{PFK}(k_A^-(k_F^-+k_{PFK})+k_F^+k_{PFK})}{k_A^+k_F^+k_{PFK}(k_a^-+k_a^+)^4}$	1	$\dfrac{k_A^+(k_F^-+k_{PFK})}{k_A^-(k_F^-+k_{PFK})+k_F^+k_{PFK}}$	$\dfrac{k_A^+k_F^+}{k_A^-(k_F^-+k_{PFK})+k_F^+k_{PFK}}$	1	1
1	$\dfrac{4(k_a^-)^3 k_a^+ v_{PFK}(k_A^-(k_F^-+k_{PFK})+k_F^+k_{PFK})}{k_A^+k_F^+k_{PFK}(k_a^-+k_a^+)^4}$	1	$\dfrac{k_A^+(k_F^-+k_{PFK})}{k_A^-(k_F^-+k_{PFK})+k_F^+k_{PFK}}$	$\dfrac{k_A^+k_F^+}{k_A^-(k_F^-+k_{PFK})+k_F^+k_{PFK}}$	$\dfrac{4k_a^+}{k_a^-}$	$\dfrac{4k_i^+}{k_i^-}$
2	$\dfrac{6(k_a^-)^2(k_a^+)^2 v_{PFK}(k_A^-(k_F^-+k_{PFK})+k_F^+k_{PFK})}{k_A^+k_F^+k_{PFK}(k_a^-+k_a^+)^4}$	1	$\dfrac{k_A^+(k_F^-+k_{PFK})}{k_A^-(k_F^-+k_{PFK})+k_F^+k_{PFK}}$	$\dfrac{k_A^+k_F^+}{k_A^-(k_F^-+k_{PFK})+k_F^+k_{PFK}}$	$\dfrac{6(k_a^+)^2}{(k_a^-)^2}$	$\dfrac{6(k_i^+)^2}{(k_i^-)^2}$
3	$\dfrac{4k_a^-(k_a^+)^3 v_{PFK}(k_A^-(k_F^-+k_{PFK})+k_F^+k_{PFK})}{k_A^+k_F^+k_{PFK}(k_a^-+k_a^+)^4}$	1	$\dfrac{k_A^+(k_F^-+k_{PFK})}{k_A^-(k_F^-+k_{PFK})+k_F^+k_{PFK}}$	$\dfrac{k_A^+k_F^+}{k_A^-(k_F^-+k_{PFK})+k_F^+k_{PFK}}$	$\dfrac{4(k_a^+)^3}{(k_a^-)^3}$	$\dfrac{4(k_i^+)^3}{(k_i^-)^3}$
4	$\dfrac{(k_a^+)^4 v_{PFK}(k_A^-(k_F^-+k_{PFK})+k_F^+k_{PFK})}{k_A^+k_F^+k_{PFK}(k_a^-+k_a^+)^4}$	1	$\dfrac{k_A^+(k_F^-+k_{PFK})}{k_A^-(k_F^-+k_{PFK})+k_F^+k_{PFK}}$	$\dfrac{k_A^+k_F^+}{k_A^-(k_F^-+k_{PFK})+k_F^+k_{PFK}}$	$\dfrac{(k_a^+)^4}{(k_a^-)^4}$	$\dfrac{(k_i^+)^4}{(k_i^-)^4}$

The total mass balance on all the forms of the enzyme,

$$\mathrm{PFK_{tot}} = \sum_{i=0}^{4}(R_i + R_{i,A} + R_{i,AF}) + \sum_{i=0}^{4} T_i, \qquad (14.22)$$

will lead to 20 algebraic equations with 20 concentration variables. This set of equations can be solved using symbolic solvers like MATHEMATICA; see Table 3.1.

In addition, the reaction rate v_{PFK} is the input flux for the reactants and the output flux for the product in the steady state. The reaction rate is

$$v_{PFK} = k_{PFK} \sum_{i=0}^{4} R_{i,AF}, \qquad (14.23)$$

where k_{PFK} is unknown. Since the binding step of the activator and the inhibitor is at equilibrium at steady state, only the corresponding equilibrium constants will appear in the steady-state equation. However, since the binding of the reactants are in a steady state, we will need numerical values for k_F^+ and k_A^+.

In principle, we can thus specify k_F^+ and k_A^+ and solve 21 equations for the 20 concentration variables and k_{PFK}, given numerical values for the total amount of enzyme and the five binding constants in Eq. (14.21). In practice, the choices for k_F^+ and k_A^+ are restricted for the computed concentrations and k_{PFK} to take on positive values.

The steady-state solution   The solution to the 19 algebraic equations with the reaction rate law (Eq. (14.23)) is given in Table 14.3. The total PFK can then be computed from Eq. (14.22).

The first column in Table 14.3 shows the solution for $R_{i,0}$. Summing up these columns gives the total amount of the $R_{0,i}$ forms. The next three columns show the relative distribution of the $R_{i,0}$, $R_{i,A}$, and $R_{i,AF}$ forms.

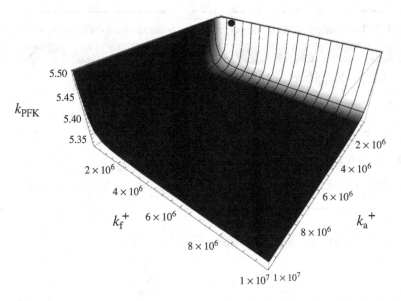

Figure 14.3 The relationships between the three key PERCs (k_a^+, k_f^+, k_{PFK}) of PFK given in Eq. (14.25). The figure is gray-scaled based on the fraction of the enzyme that is in the T form, with white indicating a high fraction in the T forms. The black dot shows the particular combination chosen.

These relative amounts are the same for all the activator bound states of the enzyme, i.e., the same for all i.

Thus, the relative amount of the enzyme in the different activator bound states is given by the relative amount of the $R_{i,0}$ forms, shown in the fourth column of the table.

The solution for T_0 is

$$T_0 = \frac{(k_i^-)^4 L v_{PFK}[k_a^-(k_f^- + k_{PFK}) + k_f^+ k_{PFK}]}{k_a^+ k_f^+ k_{PFK}(k_i^- + k_i^+ x)^4}, \tag{14.24}$$

and then the relative amount of the T_i forms is given in the last column of Table 14.3.

Numerical values We now introduce the numerical values, $K_i = 0.1/1.6$, $K_a = 0.033/0.0867$, $K_A = 0.068/1.6$, $K_F = 0.1/0.0198$, $v_{PFK} = 1.12$ mM/h, $L = 0.0011$, using the dissociation constant values given earlier in the chapter and the steady-state concentrations of the ligands. We can introduce these into the solution and sum over all the forms of the enzyme to get

$$PFK_{tot} = 1.71/k_a^+ + 1.19/k_f^+ + 7.14/k_{PFK} = 0.000\,033\,\text{mM}. \tag{14.25}$$

The three parameter values are thus not independent, as stated above. This equation can be plotted in three dimensions; Figure 14.3. For the simulations below we choose, $k_f^+ = 5.05 \times 10^7$ L/(h mM) and $k_a^+ = 1.25 \times$

10^5 L/(h mM), and compute $k_{PFK} = 3.07 \times 10^5$ L/h from Eq. (14.25). With these values, about 90% of the enzyme is in the R form ($r_R = 0.896$) and about 12% of R is in the $R_{i,AF}$ forms ($r_{cat} = 0.123$). With these values, the relative flux load through the five forms of $R_{i,AF}$ is

i	fraction
0	0.005 77
1	0.060 7
2	0.239
3	0.419
4	0.275.

$$(14.26)$$

Thus, most of the flux is carried by $R_{3,AF}$ for these parameter values and steady-state concentrations of ligands.

14.3 Integration of PFK with glycolysis

The PFK subnetwork can now be integrated with the glycolytic model of Chapter 10. The integration process is straightforward.

The stoichiometric matrix The column for the PFK reaction is replaced with the PFK subnetwork. This results in a stoichiometric matrix that is of dimensions 40×44 and of rank 37.

The null space is thus seven-dimensional ($= 44 - 37$). It fundamentally has the same three pathways as the glycolytic model alone. The difference is that pathway 1 in Figure 10.2 now becomes five pathways (p_1 through p_5 in Table 14.2), one through each of the R_i forms of PFK, consistent with the properties of the stoichiometric matrix for the PFK subnetwork. The pyruvate reduction pathway and the AMP exchange pathway remain the same as in Figure 10.2 and become p_6 and p_7 in Table 14.2, for a total of seven pathways.

The left null space is three-dimensional ($= 40 - 37$). It has the same pools as glycolysis, namely the total phosphate and NAD pools (see Table 10.7), and the total PFK pool that originates from the stoichiometric matrix of the PFK sub-network; Table 14.2.

The steady state The PFK subnetwork is set up to be in balance with its metabolic network environment. This balance is ensured by keeping the substrate and regulator concentrations the same and the total flux through the reaction the same. The steady state of the integrated model will thus be the same as for glycolysis alone. Perturbations away from the steady state will be different.

Table 14.4 The stoichiometric matrix for glycolysis with PFK. The matrix is partitioned to show the addition of PFK sub-network to the matrix for glycolysis. The pathway vectors are shown at the bottom of the table

PG13 PG3 PG2 PEP PYR LAC NAD NADH AMP ADP ATP P$_i$ H H$_2$O P$_1$ P$_2$ P$_3$ P$_4$ P$_5$ P$_6$ P$_7$

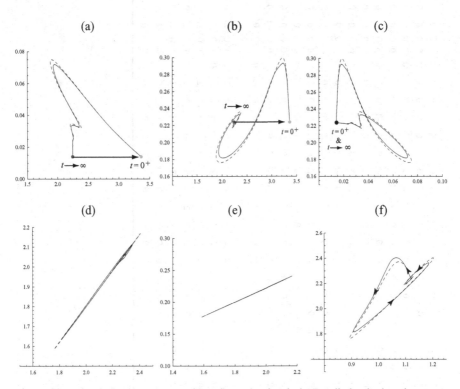

Figure 14.4 The dynamic response of key fluxes in glycolysis. Detailed pairwise phase portraits: (a) v_{ATP} versus v_{AMP}. (b) v_{ATP} versus v_{PYR}. (c) v_{AMP} versus v_{PYR}. (d) v_{GAPDH} versus v_{LDH}. (e) v_{LDH} versus v_{NADH}. (f) v_{HK} versus v_{PK}. Each panel compares the solution with (solid line) and without (dashed line) coupling of the PFK sub-network to glycolysis. The unregulated responses are the same as in Figure 10.5. The fluxes are in units of mm/h. The perturbation is reflected in the instantaneous move of the flux state from the initial steady state to an unsteady state, as indicated by the arrow placing the initial point at $t = 0^+$. The system then returns to its steady state at $t \to \infty$.

Dynamic simulation   We now perturb the integrated model by increasing the rate of ATP utilization by 50%, as in Chapter 10. We show the results in the form of the same key flux phase portraits as in Figure 10.5 for glycolysis alone; see Figure 14.4.

1. *Increase in the rate of ATP use* The overall dynamic responses to this increase in ATP utilization rate is similar with and without regulation of PFK. The enzyme in the steady state is 90% in the R form. Thus, there is little room for regulation to change. In Figure 14.5a we show the fraction of PFK in the R state versus the energy charge the response of the integrated system. The initial response to the drop in the energy charge, which results from the increased ATP usage rate, leads to an increase in PFK that is in the R state. The fraction r_R reaches almost unity, which represents the full possible regulatory response. After the response dies down, the energy

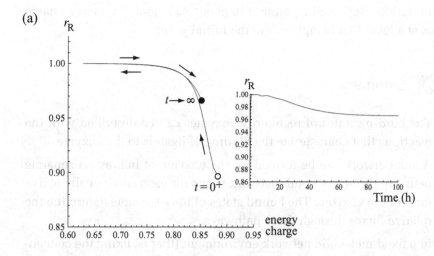

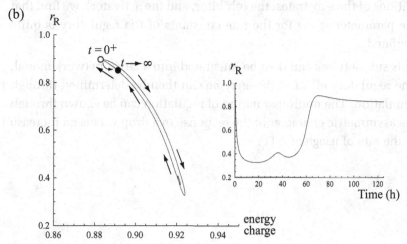

Figure 14.5 The dynamic response of key ratios. (a) The phase portrait for the energy charge versus r_R for 50% increase in ATP utilization rate (k_{ATP}). (b) The same as in (a), but with a 15% drop in k_{ATP}.

charge returns to a value close to the initial one, but the fraction of PFK in the R state, r_R, settles down at about 97%.

2. *Decrease in the rate of ATP use* The regulatory action of PFK in this steady state to an increase energy usage is thus limited. To illustrate the nonlinear nature of enzyme regulation, we now simulate the response of the integrated network to a 15% drop in the rate of use of ATP. The response is quite asymmetric to the response to an increase in ATP usage rate; Figure 14.5b. The buildup of ATP that results from the initial drop in its rate of usage rapidly inhibits PFK and the R fraction, r_R, drops below 40%. As the network then approaches its eventual steady state the ATP

concentration drops and r_R returns to about 85% and the energy charge settles at a level that is higher than the initial point.

14.4 Summary

➤ The binding states of regulatory enzymes can be described with the reactions that characterize the binding of ligands to the enzyme.

➤ A sub-network can be formed for the enzyme of interest. A separate pathway through the sub-network forms for each catalytically active form of the enzyme. The bound states of the regulators determine the relative fluxes through these pathways.

➤ In a fixed metabolic network environment (that is, fixing the concentrations of the substrates, the inhibitor, and the activator), we find that the parameter space for the rate constants of the regulatory is quite confined.

➤ This sub-network can then be integrated into a larger network model. The regulatory effect of the enzyme can then be determined through simulation. The nonlinear nature of regulation can be shown through the asymmetric character of the response to a drop versus an increase in the rate of usage of ATP.

## CHAPTER 15

Epilogue

We have come to the end of this text. The motivation for writing it is to develop the skills and procedures that are needed to build large-scale kinetic models in the era of availability of massive amounts of high-throughput data, or the so-called omics data types. Historically, kinetic models have been built based on enzymatic information obtained *in vitro*. Although, many useful dynamic models have been built this way, they are hard to scale and the kinetic constants obtained *in vitro* may not apply *in vivo* [32]. An alternative approach is now feasible. The procedures developed here are based on network reconstruction to obtain the stoichiometric matrix and the building of condition-dependent pseudo-elementary mass action kinetics on top of the network structure. A description of the challenges of large-scale model building in the omics era have been outlined [52].

15.1   Building dynamic models in the omics era

Workflow   The conceptual formulation underlying the MASS model building procedure is summarized in Figure 15.1. It represents an integration of data and mathematical analysis. The workflow that underlies these steps is outlined in Figure 15.2. Briefly, the steps involved in the process are:

1. A model describing the dynamic states of a network requires us to first establish the network to be studied. In the pre-genome era, network reconstruction was based on assembling biochemical data from a variety of sources, e.g., see [107, 120]. Since whole-genome sequencing emerged [121], genome-scale network reconstruction has become possible [29, 30]. There are now a variety of sources to aid with the

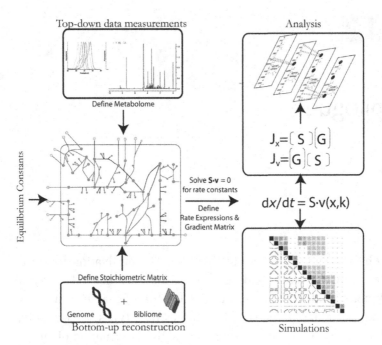

Figure 15.1 Conceptual basis for the formulation of MASS models. Modified from [54]. This figure parallels Figure 1.6.

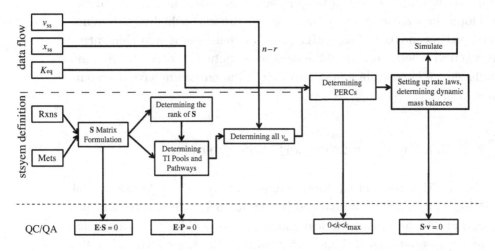

Figure 15.2 Workflows for the formulation of MASS models. Prepared by Aarash Bordbar.

reconstruction process. The network reconstruction process has been conceptually described [100], the workflows have been laid out [35], and the standard operating procedures detailed [117].

2. Omic data is then mapped into the reconstruction. After defining the boundaries of the system and the dimension of the null space, uptake and secretion rates (i.e., the I/O) and any fluxomic measurements are

used to estimate the steady-state flux map. The pathway basis for the null space can then be used to interpret this state.

3. The omic data-mapping process involves the use of metabolomics data, proteomics data, and targeted abundance measurements. This data, in conjunction with the steady-state flux map, allows us to estimate the rate constants *in vivo*. The rate constants are viewed as PERCs. These are condition dependent, representing the condition under which the data was obtained. Any known equilibrium constants reduce the dimensionality of the parameter estimation process, as such constants should relate two PERCs. Any information about the genetic basis for altered protein–protein or protein–ligand binding constants introduces genetic information into the model constructed.

4. The formulated dynamic mass balances can then be analyzed and simulated; recall Figure 1.6. Dynamic simulation is straightforward to carry out. Numerical problems may be encountered when the system has a very wide range of time constants. The logarithm of the ratio between the largest and that smallest eigenvalue of the system is known as the condition number. Systems with large condition numbers (e.g., larger than 8 to 10) may become numerically stiff and hard to solve. In the past, the QSSA and QEA were used to enable analytical and numerical tractability. In today's environment, numerical solutions are easier to get. Detailed analytical solutions at a large scale, even if achievable, can be replaced by easy-to-understand visualization of solutions.

5. The interpretation of the result of the simulation comes down to insightful graphing of the data. Biochemical and time-scale separation analysis leads to the formation of aggregate variables, or pools, that systematically form over multiple time scales. The ratios of these pools often form quantities that are physiologically meaningful and can be used to convert network maps (that are biochemical in nature) to operating-type diagrams (that are more systems engineering in nature). This latter representation leads to a systems thinking and representation of how a network actually carries out its physiological functions. The analysis methods associated with MASS models are summarized in Figure 15.3.

Issues and challenges   The MASS procedure works well in principle and is readily scalable. As with any modeling procedure, there are challenges that grow with scale and the quality of the information used to build the models. In fact, one of Bailey's reasons for model building (Section 1.2) is to interrelate data types and examine their consistency.

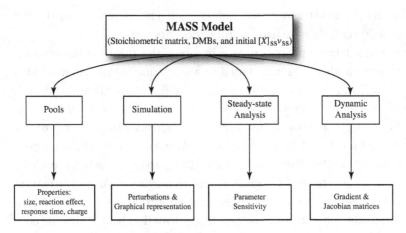

Figure 15.3 Analysis methods associated with MASS models. Prepared by Aarash Bordbar.

With the MASS model construction process comes new issues and challenges. Four are highlighted here:

1. *Data completeness*: although the omics data types strive to be comprehensive or even complete, there will be missing measurements. For instance, some metabolites are hard to detect or they can be present in very low amounts. Procedures on how to deal with incomplete data sets will need to developed.

2. *Data quality/error*: the information that goes into a model determines its quality and use. Traditional methods for parametric sensitivity analysis exist and they need to be deployed to relate the error in the data used to build a model and how it affects the interpretation of dynamical states of the underlying network. Their use to design critical experiments should not be overlooked.

3. *Condition specificity*: unlike biophysical models that are based on first principles and thus have "absolute" characteristics, MASS models will always be "relative" to the condition under which the omics data is obtained. The network structure can have finality to it, but its functional states are condition dependent.

4. *Validation and prospective uses*: for many types of model, validation procedures have been developed. Such procedures for MASS models need to be established. The predictive uses of large-scale dynamic models in cell and molecular biology represent an interesting challenge and an exciting future prospect.

Although these are research challenges for the future development of MASS models, the reader and student of this book will have learned how to build and simulate large-scale dynamic models. These lessons are not specific to MASS models, but apply to dynamic simulation in general.

MASS versus classically built kinetic models   The traditional approach to building large dynamic models of metabolic networks basically represents scaling of *in vitro* enzyme kinetic data to the treatment of multiple enzymes functioning simultaneously inside the living cell. These dynamic models are biophysical in nature. As outlined above, advancements in high-throughput technologies have driven the formulation of MASS models that in turn present a class of models that can be built through utilization of omics data. MASS models are a formal way to integrate omics data sets.

In vivo MASS models provide an alternative to the traditional *in vitro* data-based kinetic models; both account for regulation and exhibit biological properties such as time-scale decomposition. MASS models have the added benefits of incorporating *in vivo* data, scalability, and direct representation of the regulatory interactions. Some of the advantages and disadvantages of the two approaches are detailed in Table 15.1. Each of the approaches has challenges related to data measurement errors and data completeness; however, the mathematical formulation, algorithmic construction, and omic data reliance of MASS models make these models more amenable to constraint-based analysis methods; see the companion book [89].

Level of detail   One issue that comes up with scale is how detailed does a description have to be. This question will be especially pertinent when incorporating proteins that can have many ligands bound and many functional states. In addition, more and more information about the phosphorylated state of proteins is becoming available and phosphorylated states can be measured through phospho-proteomics. Such proteins act as integrators of many different signals.

In the omics era we have a paradigm to drive towards completeness. The debate in the early days of the human genome project was an early indication of the differing opinions of the resolution to which the data should be obtained. In the ensuing years, it has become clear that we can generate the data with the high-throughput technologies. The question is really one of analysis and to what level of detail one needs to describe a particular system. Given the multiscale nature of biological systems and data sets, the answer to this question will be left to the user. When constraints-based reconstruction and analysis (COBRA) methods reached the genome scale, it became clear that having all the information available

Table 15.1 Comparison of the traditional and the MASS approach to building large-scale kinetic models. From [54]

	Traditional models	MASS models
Condition specificity	None (global); aim to account for many details, including temperature and pH dependence	Tailored; all condition-dependent factors/parameters are lumped into condition-specific constants
Protein activity resolution	Implicit through quasi-steady state and quasi-equilibrium assumptions; details about protein–protein interactions and intermediates are often masked	Explicit stoichiometric mass action representation; allows direct integration as well as interrogation of proteomic and metabolomic data from an integrated network perspective
Model building and formulation	Case-by-case treatment of enzymes and their kinetics	Algorithmic approach based on omics data
Mathematical characteristics	Hyperbolic equations	Bilinear equations
Scalability	Poor due to the case-by-case treatment of additional enzymes	Easy, but has a larger number of variables
Data errors (rate constants)	Stymied by the recognized differences in *in vitro* versus *in vivo* kinetics	Calculated from concentrations and subject to errors in these measurements. Assessment of experimental errors explicit
Data completeness	Limited by ability to characterize all enzymes *in vivo* and measure all metabolites *in vivo*	Limited by the coverage of omics datasets. Procedures to deal with incomplete measurements need to be developed
Validation	Mature, well developed; pros and cons relatively well known	At the early stages of development and more experience is needed at large scale

was important, and a particular user could filter the information to suit a particular purpose [36, 86].

15.2   Going forward

Education   Historically, the overwhelming complexity of biological systems has made it difficult to uncover basic underlying mechanisms. However, at the same time, the need to deal with integrated, evolving biological systems in the absence of such knowledge has made for descriptive, metaphorical, and historical approaches. Given this background, it is not surprising that relatively little has been done to integrate the findings of

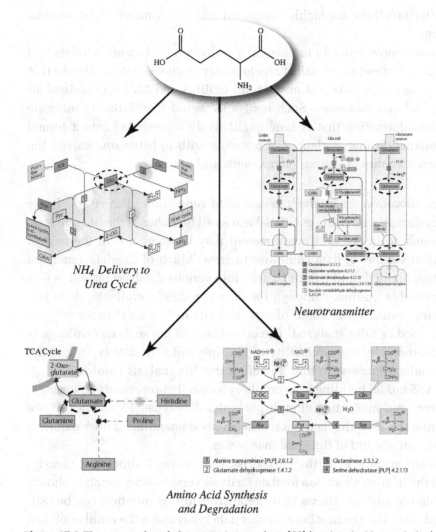

Figure 15.4 The many roles of glutamate. Images from [68] (reprinted with permission).

molecular and cellular biology to achieve an integrated understanding of intact biological systems.

Biology education reflects this character. Biology courses tend to be descriptive. They emphasize a historical approach that stresses key experiments, particular model systems, and the logic of experimental design. The data is disseminated into a logical progression of chapters. Biochemistry courses are an example. The student learns topic after topic. When preparing for the first exam the student may be faced with the situation illustrated in Figure 15.4. The same compound appears in many chapters in the book and is described to have a different role in each one. However, in reality, the molecule is simultaneously involved in many processes.

Cellular functions are highly integrated and coordinated. Enter systems biology.

It seems now possible to develop a collection of dynamic models that will help convert the classical biochemistry textbooks into textbooks that will have toy models and meaningful COBRA and MASS models of all major cellular processes. Such textbooks would then formally integrate all the information that is now qualitatively represented into a formal representation that students can compute with to better understand the interrelationship of various components and processes.

Effort allocation for analysis versus data generation   Larger and larger omics data sets are being generated across all branches of the life sciences. Although costs per unit measurement may be going down, the cost of the total volume of data continues to grow. Much of the data generated through public funding finds its way into various database resources that are available on-line. Although the data is publicly available, there is a growing concern that much of this data sits in these databases without being used or fully analyzed. The complexity of the analysis challenge is compounded by the availability of multiple different data types and the concomitant increase in time and effort that the analysis requires.

MASS model building is a complex process that represents an example of structured integration of multiple omics data types; Figure 15.1. It is a laborious process that is quite intellectually demanding. It does require a significant amount of time and manpower.

These concerns raise the critical issue of resource allocation; namely, what the relative allocation for data analysis versus data generation should be. Simple analysis shows that the answer to this question can be estimated. Let m be the number of omics data types and n the number of data sets or measurements made; i.e., m would be 2 for integrated analysis of transcriptomic and proteomic data. Then we have that the effort for data generation is

$$\text{generation} \sim nm \tag{15.1}$$

and the effort for data analysis is

$$\text{analysis} \sim n^m. \tag{15.2}$$

Let a be the effective dimensionality of the data set, say as determined by singular value decomposition, relative to n. The fraction of the total effort that is required for data analysis is then

$$f = \frac{\text{analysis}}{\text{generation} + \text{analysis}} = \frac{an^m}{an^m + nm}. \tag{15.3}$$

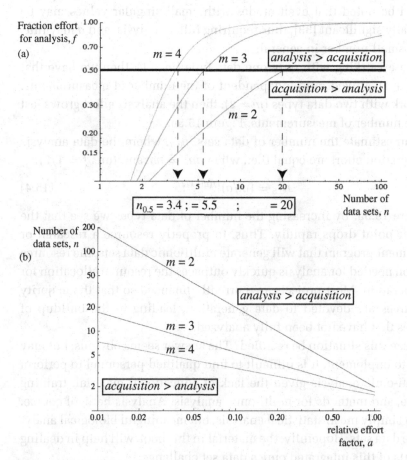

Figure 15.5 Estimation of the relative resource allocation for experiment and for analysis. (a) The fraction f of resource allocation for analysis as a function of the number of data sets n; $m = 3$ and $a = 1, 0.1, 0.01$. (b) The equi-effort number of experiments ($n_{0.5}$) as a function of $m = 2, 3, 4$.

The constant a is a scaling factor that measures three things:

- First, the intrinsic ratio of effort for experiments versus data (i.e., it may take a day to do the experiment and it may take an hour to analyze the data).
- Second, the dimensionality of the data set. If the intrinsic ratio is roughly unity and we need 10 measurements to determine one independent piece of information, then $a = 0.1$.
- Third, since all m data types may not take exactly the same effort, nm needs to be scaled accordingly. This scaling factors into a when f is evaluated.

The dimensionality of omics data sets is relatively low, as pointed out in an early expression profiling data set throughout the yeast cell cycle [12].

It should be noted that even modes with small singular values may be biologically significant [33], thus requiring fuller analysis, and a may thus not be a small number in general.

If we are working with only one data type ($m = 1$), then we have that $f = a/(a + 1)$, a fixed ratio independent of the number of measurements. If we work with two data types ($m = 2$), then the analysis effort grows fast with the number of measurements; Figure 15.5a.

We can estimate the number of data sets, $n_{0.5}$, where the data analysis and generation effort are equal (i.e., when $an^m = nm$ and thus $f = 1/2$):

$$n_{0.5} = (m/a)^{1/(m-1)};\qquad(15.4)$$

see Figure 15.5a. By increasing the number of data types we see that the crossover point drops rapidly. Thus, to properly resource a research or development program that will generate multi-omic data sets, the resource allocation needed for analysis quickly outpaces the resource allocation for data generation. Most projects are currently financed so that the majority of resources are devoted to data generation, leading to the buildup of databases that have not been fully analyzed.

How can this situation be rectified? The answer seems obvious, but may be hard to implement. It is difficult to find qualified personnel to perform the multi-omic analysis given the lack of educational material, training available, and methods for multi-omic analysis. Analysis here, of course, does not simply mean statistical analysis, but meaningful biological analysis of the data sets. Hopefully, the material in this book will help in dealing with parts of this integrated omics data set challenge.

Nomenclature

Symbols

b_i	individual exchange fluxes
$\mathbf{e}_i$	a row vector in the elemental matrix giving the elemental composition of compound i
$\mathbf{E}$	the elemental matrix giving the composition of all the compounds considered in a network. The columns correspond to the compounds and the rows to the elements
f_i	a fraction of molecules in a particular stage
$\mathbf{g}_i$	a row in the gradient matrix
g_{ij}	an element in the gradient matrix
$\mathbf{G}$	the gradient matrix
$\mathbf{H}$	the Hessian matrix
$\mathbf{I}$	the identity matrix
$\mathbf{J}$	the Jacobian matrix
$\mathbf{k}$	the vector of kinetic parameters
k_i	rate constant for reaction i; also pseudo-elementary rate constant (PERC)
$\mathbf{K}$	a lumped parameter from the rate constants of a reaction mechanism
$\mathbf{K}_{eq}$	equilibrium constant of a reaction
$\mathbf{K}_m$	the Michaelis–Menten constant
$\mathbf{L}$	a matrix of left null space basis vectors
$\mathbf{l}_i$	a left null space basis vector, a row vector
$\mathbf{P}$	a pool transformation matrix
$\mathbf{p}$	a vector of pooled variables
p_i	size of conservation pools, i, in units of concentration

$\mathbf{p}'$	a vector of pooled deviation variables
r	common symbol for the rank of a matrix; also a fraction of molecules
$\mathbf{r}_i$	a right null space basis vector, a column vector
$\mathbf{S}$	the stoichiometric matrix, each row corresponds to a metabolite and every column to a reaction. Dimension given in terms of $m \times n$.
$\mathbf{s}_i$	the reaction vector corresponding to reaction i, a column vector of the stoichiometric matrix
s_{ij}	an element in the stoichiometric matrix
t	used for time
$\mathbf{u}_i$	eigenrows
$\mathbf{V}$	matrix of right singular vectors
$\mathbf{v}$	the vector of reaction fluxes, dimension is n. Also an eigenvector
v_i	the flux through the ith reaction; units are moles per volume per time. Also v as the number of molecules of an inhibitor, I. Also the degree of cooperativity in the Hill model
$\mathbf{w}$	the vector of weights
w_i	element i in a vector of weights
$\mathbf{x}$	the vector of concentrations, dimension is m.
$\mathbf{x}'$	the vector of concentration deviation variables
x_i	the concentration of the ith compound; units are moles per volume
z_i	the charge of a molecule
Z	an objective function, a scalar

Greek symbols

Γ	a mass action ratio
Δ	a characteristic change in a state variable
$\Delta\mu_{H^+}$	protonmotive force
ΔpH	hydrogen ion gradient
$\Delta\Psi$	charge gradient
λ_i	eigenvalues; also a net rate constant
π_i	dimensionless multipliers; also participation number
Π_i	used to define osmotic pressure; also indicates products of concentrations

ρ_i	connectivity number
Σ	diagonal matrix of singular values; also a summation of components
τ	general symbol for a time constant
χ	a dimensionless concentration

Mathematical symbols

Col(**A**)	column space of matrix **A**
Left Null(**A**)	left-null space of matrix **A**
Null(**A**)	null space of matrix **A**
Row(**A**)	row space of matrix **A**
T	transpose
$\| \cdot \|$	the norm of a vector
$\langle u, v \rangle$	inner product
$\langle u \oplus v \rangle$	outer product

Homework problems

B.1 Introduction

1.1 Read Part 1 of [89] to familiarize yourself with the process of reconstructing biochemical reaction networks. These are the networks whose dynamic states we want to study.

1.2 Read Chapters 6, 9, and 10 in [89] to familiarize yourself with the stoichiometric matrix and its null spaces.

1.3 Take a couple of hours to browse the websites in Table 1.1

B.2 Basic concepts

2.1 Derive Eqs (2.14), (2.15), and (2.16).

2.2 For a reaction that has a net flux of 1, display graphically how $\Gamma \to K_{eq}$ as the forward reaction rate increases from 1 to 1000.

2.3 Derive the net reaction for

$$x_1 + x_2 \rightleftharpoons x_3 + x_4 \tag{B.1}$$

using mass action kinetics. Show the rate expression when $(1 - \Gamma/K_{eq})$ is considered as a factor in the equation.

2.4 Formulate the details of

$$\text{charge} = \frac{\text{occupancy}}{\text{capacity}} \tag{B.2}$$

for three cofactors in Table 8.1, including CoA.

B.3 Dynamic simulation: the basic procedure

3.1 Let us consider that a chemical system is described by the concentration of 10 variables denoted by $x_1, x_2, \ldots, x_{10}$.

   (i) By assigning numerical random values between 0 and 1 to the 10 variables, construct a hypothetical time series comprised of 100 measurements at different times.

  (ii) Verify that in this case the time correlation matrix $\mathbf{R}$, defined in Eq. (3.7), shows no correlation.

3.2 Add a fourth pool to Eq. (3.5) that represents the total number of phosphate bonds.

3.3 Consider the reaction schema of Eq (3.9). Change the equilibrium constants in Eq. (3.12) to $K_1 = K_2 = 2$.

- Redo the simulations on the time scales shown in Figure 3.7.
- Compute the correlation coefficients on the three dominant time scales.
- Suggest a pool forming matrix $\mathbf{P}$.
- Post-process $\mathbf{x}(t)$ with your suggested $\mathbf{P}$ and plot the phase portraits for the pools. Did your pool formation matrix generate dynamically independent pools as in Figure 3.9?

3.4 Consider the reaction schema (3.9) with $k_1 = 1.0$, $k_2 = 0.0001$, $k_3 = 0.01$, with $K_1 = K_2 = 1$. Find a pool forming matrix $\mathbf{P}$ that represents dynamically independent pools.

3.5 Consider one additional reaction to reaction schema (3.9)

$$X_1 \underset{}{\overset{v_1}{\rightleftharpoons}} X_2 \underset{}{\overset{v_2}{\rightleftharpoons}} X_3 \underset{}{\overset{v_3}{\rightleftharpoons}} X_4 \overset{v_4}{\rightarrow} X_5$$

with reaction rates

$$v_1 = k_1(x_1 - x_2/K_1), \quad v_2 = k_2(x_2 - x_3/K_2)$$

$$v_3 = k_3(x_3 - x_4/K_3), \quad v_4 = k_4 x_4.$$

The parameter values that we will use in this example are $k_1 = 1.0$, $k_2 = 0.01$, $k_3 = 0.0001$, and $k_4 = 0.000\,001$ with $K_1 = K_2 = K_3 = 1$.

- Redo the analysis performed in the text with the one additional reaction.
- What is the appropriate form for $\mathbf{P}$ to get dynamically independent pools?
- Can you generalize this result to $5, 6, 7, \ldots, n$ reversible linear reactions in a series where all the equilibrium constants are unity?
- What happens if the reactions are irreversible ($K \to \infty$)?

B.4   Chemical reactions

4.1 Consider the reactions in Eq. (4.32).

(i) Assign numerical values to $k_1 = k_{-1} = k_3 = k_{-3} = 1$ and $k_2 = 1, 0.1, 0.01$. Simulate the dynamics of this system for the three cases. Interpret the dynamics in terms of the pools and the time-scale separation that occurs at low values for k_2.

(ii) Show that $p_2 + p_4$ is a constant, which leads to the formation of the last line in Eq. (4.41).

(iii) Use $k_1 = k_{-1} = k_2 = 1$ and vary $k_3 = 0.01, 1, 100$ with $k_3 = k_{-3}$. Show that the dynamics of reaction 1 do not change. Use $x_1(0) = 1$, $x_2(0) = x_3(0) = x_4(0) = 0$ for all simulations.

4.2 Consider the reaction in Eq. (4.10).

(i) Derive Eq. (4.17).

(ii) Derive Eq. (4.20).

(iii) Re-derive these equations for $x_1(0) = 3$, $x_2(0) = 4$, and numerically simulate the dynamics of the reaction and produce a plot analogous to Figure 4.2.

4.3 Equation (4.12) represents the disequilibrium pool for a bilinear reaction. Verify the following:

(i) Combining Eqs (4.15) and (4.16) obtains Eq. (4.17).

(ii) Combining Eqs (4.15) and (4.16) by using $x_1(0)$ instead of 3 and $x_2(0)$ instead of 2 obtains the general form of Eq. (4.17). Find the general solution for the equilibrium concentrations.

(iii) Through a Taylor series expansion around a steady state $x_{eq} = (x_{1,eq}, x_{2,eq}, x_{3,eq})$, verify the linearized form of Eq. (4.13) is (4.20). Use Taylor expansion and show that the approximation in general is given by

$$\frac{dx_1}{dt} = k_1 x_{2,eq}(x_1 - x_{1,eq}) + k_1 x_{1,eq}(x_2 - x_{2,eq}) - k_{-1}(x_3 - x_{3,eq}).$$

4.4 Equation (4.24) allows us to represent the dynamics of the single reversible bi-linear reaction in terms of pooled variables. Verify that the coefficient relating the time derivative of the pool to itself, Eq. (4.27), is the first norm of the gradient matrix, $(k_1 x_2 + k_1 x_1 + k_{-1})$, using the results from the previous problem.

4.5 Recast the stoichiometric matrix and the reaction vector in Eq. (4.33) in terms of the net reactions. The stoichiometric matrix will have dimensions of (4×3) and the reaction vector a dimension of 3. Verify that the left null space does not change. Does the left null space change if the reactions are irreversible?

4.6 In Section 4.5, the intermediary metabolite x_3 established the dynamic coupling between the two bilinear reactions. Considering that the systems reached equilibrium, determine how x_3 depends on the pools defined in Eqs (4.42)–(4.45). What is the order of the polynomial equation for x_3? In the particular case when $p_3 = 1$, $p_4 = 2$, and $p_5 = 3$, show that x_3 can have four possible solutions. **Hint:** Take into account that at the steady state $p_1 = p_2 = 0$. Then, solve the set of Eqs (4.42)–(4.45) for all x_3. We suggest using the *Solve* command in Mathematica.

4.7 Dynamic decoupling in biochemical systems can result from network topology. A clear example is provided by the two connected reversible linear reactions treated in Section 4.4. Consider numerical values of $k_1 = k_{-1} = 1$ and $k_2 = 0.2$ and randomly select five numerical values in a range from 0 to 100 for k_3 and k_{-3} and simulate the response of the five systems to the same initial condition. Plot the results for x_1 and x_2 against time. How do you explain the results? Now graph x_3 and x_4 in the same way. **Hint:** Use the workbook example 4.2.3 and the *Random* command in Mathematica. Initial conditions are $x_1 = 1$, $x_2 = x_3 = x_4 = 0$.

4.8 For the example of the two connected reversible reactions, consider the situation in which $k_1 = k_{-1} = k_3 = k_{-3} = 1$ and $k_2 = 0.2$ with the same initial condition specified in the text. Verify that the rows for the dynamic equation for the pools

$$\frac{dp}{dt} = \begin{pmatrix} -(k_1 + k_{-1})p_1 + \frac{k_2}{K_1}p_2 \\ k_1 p_1 - k_2 p_2 \\ -(k_3 + k_{-3})p_3 + k_2 p_2 \\ 0 \end{pmatrix} \tag{B.3}$$

have the following interpretation:

(i) Last row indicates the conserved metabolites in the reaction at any time scale.

(ii) First and third rows represent the disequilibrium quantities (the first has a slow drift) which tends to be zero when the system reaches equilibrium.

(iii) Pool p_2 has the same behavior as the variable x_2.

4.9 Consider the reaction system of Eq. (6.1).

(i) Apply the QSSA so that x_2 can be computed instantly as a function of x_1.

(ii) Simplify the two-dimensional dynamic system to one that is one-dimensional.

(iii) What are the initial conditions for the one-dimensional system? Discuss the implications of the QSSA.

(iv) Redo the simulation in Figure 4.7 for the one- and two-dimensional models.

(v) Plot both solutions and compare them.

B.5   Enzyme kinetics

5.1 The application of the QSSA and QEA are two ways to simplify the solution of the mass action model for the Michaelis–Menten reaction mechanism.

(i) Verify that the dynamic behavior of the substrate in a QSSA is given by

$$\frac{ds}{dt} = \frac{-k_2 e_0 s}{K_m + s},$$ (B.4)

where $K_m = (k_{-1} + k_2)/k_1$ is the Michaelis–Menten constant. **Hint**: consider that $dx/dt = 0$.

(ii) Verify that under the QEA the differential equation for the substrate is given by

$$\frac{ds}{dt} = \frac{-k_2 e_0 s}{K_d + s + \frac{e_0 K_d}{K_d + s}}$$ (B.5)

where $K_d = k_{-1}/k_1$. **Hint**: Take into account the mass conservation and the fact that the quasi-equilibrium state is valid when $ds/dt = 0$.

(iii) Derive the differential equation for the products p considering QSSA and QEA. What are the differences between them? What are the limits of the parameter values at which the solutions converge?

(iv) Considering the QEA and that $s_0/K_m = 1$ and $e_0/K_m = 1$, prepare the phase plane of the substrate and the product using the following parameter values:

(1) $k_2/k_{-1} = 0.01$;

(2) $k_2/k_{-1} = 0.1$; and

(3) $k_2/k_{-1} = 1.0$.

Graph all solutions on the same plot.

How do these phase planes compare with the exact numerical solution? Can we conclude that k_2/k_{-1} is the primary determinant on the applicability of the QEA ($k_2 \ll k_{-1}$)?

5.2 A simple extension of the Michaelis–Menten reaction mechanism to account for commonly observed cooperativity is the two-site dimer [93]:

$$S + E \overset{k_1}{\underset{k_{-1}}{\rightleftharpoons}} X \overset{k_2}{\rightarrow} E + P$$

$$S + X \overset{k'_1}{\underset{k'_{-1}}{\rightleftharpoons}} Y \overset{k'_2}{\rightarrow} X + P$$

(B.6)

Develop a rate law following the five steps outlined in Section 5.2.

5.3 The Hill model can also be used to describe enzyme activation of the enzyme E to an active form X through the binding of ν molecules of an activator A. Here, X can catalyze the same reaction as E:

$$S + X \overset{k'}{\longrightarrow} X + P \, ,$$

(B.7)

but at a different rate, as characterized by $k' > k$.

(i) Show that the rate law now becomes

$$v(a) = v_m \frac{1 + \alpha(a/K_a)^\nu}{1 + (a/K_a)^\nu}, \quad \text{where} \quad \alpha = \frac{k'}{k} = \frac{t_{\text{turnover}}}{t_{\text{activation}}}.$$

(B.8)

(ii) Show that the sensitivity of the reaction rate with respect to a is

$$v_a = \frac{(\alpha - 1)\nu v_m}{a} \frac{(a/K_a)^\nu}{[1 + (a/K_a)^\nu]^2};$$

(B.9)

that is, $(\alpha - 1)$ times that in Eq. (5.25).

5.4 The allosteric mechanism can account for kinetic effects of the substrate s in a similar fashion as it does for the inhibitor.

(i) Show that if the substrate is allowed to bind sequentially to the E conformational state of the enzyme, one obtains a rate law:

$$v(s, i) = v_m \frac{(s/K_s)(1 + s/K_s)^{\nu-1}}{(1 + s/K_s)^\nu + L(1 + i/K_i)^\nu};$$

(B.10)

and if an activator a can bind to the E form serially, show that

$$v(s, i, a) = v_m \frac{(s/K_s)(1 + s/K_s)^{\nu-1}}{(1 + s/K_s)^\nu + L(1 + i/K_i)^\nu/(1 + a/K_a)^\nu},$$

(B.11)

where K_s and K_a are the dissociation constants for the substrate and activator respectively. The activator and inhibitor thus act to perturb the natural equilibrium between E and X by modifying L as

$$L' = L \left(\frac{1 + i/K_i}{1 + a/K_a} \right)^\nu.$$

(B.12)

(ii) Verify that the control properties of these rate laws parallel those of the simpler Hill rate law.

5.5 Derive the symmetry rate law (Eq. (5.51)) for $\nu = 2$ and $\nu = 3$.

5.6 From the general symmetric model described in Section 5.5, consider the number of binding sites to be 2 ($\nu = 2$). For the particular situation where the rate constants are $k^+ = k^- = k_i^+ = k_i^- = 100$ and $k = 1$, and where the initial conditions are $s(0) = 10$, $e(0) = 5$, $p = 0$, $x = 0$, $i(0) = 10$, $x_1 = 0$ and $x_2 = 0$:

(i) Construct the stoichiometric matrix and use it to verify that there are three conservative quantities.

(ii) Compute and plot:

(a) the phase plane of s and e;

(b) the phase plane of i and a variable defined as the sum of $x_1 + x_2$;

(c) the phase plane of i and a variable defined by the sum of $x_1 + 2x_2$;

(d) the fast and slow transients for all the variables.

5.7 Consider the reaction schema (3.9) with $k_1 = 1.0$, $k_2 = 0.01$, $k_3 = 0.0001$, with $K_1 = K_2 = 1$ and initial conditions $x_1 = 1$ and $x_2 = x_3 = x_4 = 0$. Use the pool forming matrix $\mathbf{P}$

$$\mathbf{P} = \begin{pmatrix} 1 & -1 & 0 \\ 0.5 & 0.5 & -1 \\ 1 & 1 & 1 \end{pmatrix}$$

that represents approximately dynamically independent pools to:

(i) Find a dynamic description of the faster time scale by considering the two slowest pools at the initial condition.

(ii) Find a dynamic description of the middle time scale by considering fast pools at quasi-equilibrium and the slowest pool at its initial condition.

(iii) Find a dynamic description of the slowest time scale by considering two fastest pools at quasi-equilibrium.

(iv) Examine the appropriateness of your simplified descriptions using dynamic simulation of the full set of equations and your simplified ones on the appropriate time scale. Plot both the full and the simplified solutions together on the same graph.

5.8 In Section 5.4 we illustrated how to estimate the time constant for a Hill-type equation with a degree of cooperativity of ν. This estimation was based on determining the inflection point in the rate law. Verify that the symmetric model has a similar characteristic:

(i) The sensitivity of the reaction rate to the end product concentration is given by

$$v_i = \frac{\partial v}{\partial i} = -\frac{(i + K_i)^{\nu-1} L v_m \nu}{[1 + L(1 + i/K_i)^\nu]^2}. \tag{B.13}$$

(ii) Verify that the evaluation of v_i at the inflection point (i^*) is

$$v_i^* = -\frac{V_m}{K_i}N(v), \tag{B.14}$$

where

$$N(v) = L^{1/v}\frac{(v-1)^{1-1/v}(1+v)^{1+1/v}}{4v}. \tag{B.15}$$

(iii) Determine time constant that characterizes the inhibitory process represented by Eq. (5.51).

B.6 Open systems

6.1 Can you think of the following as systems: (1) DNA (a plasmid, a genome, or a chromosome), (2) a biological process (adaptation, migration, etc.), (3) a pathway, (4) a functional module (such as photosynthesis), (5) an organ, (6) a disease, or (7) microbial symbiosis. If so, describe the system boundary, the inputs and output, and the other attributes discussed in Section 6.1.

6.2 The reversible reaction in an open setting (Eq. (6.1)).
   (i) Compute the mass action ratio Γ/K_{eq} from Eq. (6.13) for the simulation in Figure 6.4. What do you conclude from your results.
   (ii) Redo the simulation, changing b_1 from 0.01 to 0.005 at time zero.
   (iii) Redo computations from part (i). How does the system move relative to equilibrium?

6.3 Compute $[x_{ss}/(s_{ss}e_{ss})]/[x_{eq}/(s_{eq}e_{eq})] = \Gamma/K_{eq}$ for the Michaelis–Menten reaction mechanism operating in an open environment. What are the values of Γ/K_{eq} when (each) one of the parameters is significantly larger than the others? When the parameters are all equal?

6.4 In Figures 6.3 and 6.4, pool p_1 is a disequilibrium pool and p_2 is a conservation pool for Eq. (6.1), as described in the text following Eq. (4.9). Similar pools can be defined and simulations performed for the Michaelis–Menten mechanism in an open environment.
   (i) Define expressions for the disequilibrium pool p_1 and the time-varying substrate conservation pool p_2.
   (ii) Using rate constants $k_1 = k_{-1} = 100$, $k_2 = 10$, $k_3 = 1$, and $b_1 = 0.1$, and initial conditions $s = e = 1$ and $x = p = 0$, simulate the pools until steady state is reached and provide both a phase plane and concentration–time plot, as in Figure 6.3. Explain the motion of the pools.
   (iii) Starting with the rate constants and steady-state values from (ii), perturb the system by changing $b_1 = 0.2$ at time zero. Simulate

until a new steady state is reached and provide graphs as in Figure 6.3b. Explain the motion of the pools.

6.5 Use the rate constants in Figure 6.6 with $b_1 = 0.025$ to find the steady-state values for the metabolites in the Michaelis–Menten mechanism in an open environment. For each of the following perturbations, use these steady-state values as your initial concentrations, replicate the simulations and plots in Figure 6.6, and discuss the results.

(i) Change $b_1 = 0.045$ at time zero.

(ii) Change $b_1 = 0.050$ at time zero.

(iii) Change $b_1 = 0.055$ at time zero.

B.7   Orders of magnitude

7.1 Intracellular pH and proton numbers. At pH 7, $[H^+] = 0.1$ μM. Compute the number of protons per cubic micrometer. What is the difference between the proton count at pH 7.5 and 7.8? With an intracellular volume of about a cubic micrometer and translocation of protons across membranes into the periplasmic space rapidly, can you conceptualize the dynamics of proton trafficking in cells?

7.2 Potential gradients across membranes. The energy production density in cells is very high, as estimated above. There are additional interesting aspects of the cellular energy-generating process. As stated above, the potential gradient across energy-transducing membranes is about -220 to -240 mV. The thickness of a bilipid layer membrane is about 7 nm. Compute the potential gradient in V/cm. Compare your number with the fact that the ignition coil in the car can produce 1000 V/cm potential gradient that causes sparks in air [5].

7.3 In the design of a bioreactor, it is often desirable to do back-of-the-envelope calculations prior to implementation. A photosynthetic bioreactor was proposed [57] which would depend on the growth of *C. vulgaris*, which has a cellular volume of 30 fL (1 fL $= 10^{-15}$ L) and a photosynthetic efficiency of ~23%. Cells were grown under a light with an intensity of 1.7 mW/cm^2, with 30–35% of the light in the frequency appropriate for photosynthesis. The bioreactor had a specific surface area of 3.2 cm^2/cm^3. During operation, the bioreactor was suited to grow up to 10^9 cells/mL solution. Assuming a Z scheme photosynthetic mechanism of eight photons per molecule of O_2 produced and *Chlorella* biomass enthalpy of ~25 kJ/g dry weight and that each cell has a dry/wet weight ratio of 0.25, calculate the predicted rate of cell growth (in mg dry weight per liter per hour) and the rate of oxygen production (mmol/(L h)).

7.4 The concentration of human biphosphoglycerate mutase (BPGM) was found to be nearly 60 μg/mL in erythrocytes. Given the BPGM molecular mass of 60 kDa and that total protein concentration of an average erythrocyte is 30 g per 100 ml, estimate the fraction of BPGM enzymes in human erythrocytes.

7.5 The average human erythrocyte is 6–8 μm in diameter. Estimate the range of diffusional response times for low molecular weight metabolites in the erythrocyte.

7.6 Human blood contains 160 g of hemoglobin per liter of blood. Blood contains about 5×10^9 red blood cells per milliliter. Although each red blood cell is a biconcave disk, to simplify calculations, we may regard the cells as simple cylinders with the following dimensions:

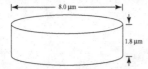

(a) Calculate the weight of hemoglobin in one red blood cell.
(b) Calculate the number of molecules of hemoglobin in one red blood cell.
(c) Calculate the volume of the red blood cell.
(d) Hemoglobin is a globular protein 6.8 nm in diameter. What fraction of the total volume of the red blood cell is occupied by hemoglobin?
(e) The ratio of the total volume of hemoglobin to the total volume of the red blood cell (part (d)) gives a deceptive picture of how tightly hemoglobin is packed in the red blood cell. One must remember that when spheres are packed, the voids between the spheres occupy a substantial portion of the total volume. Assume that the hemoglobin is packed in the red cell in a cubic array, as indicated in the following figure:

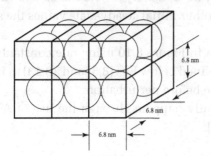

Calculate the total volume of the array of hemoglobin molecules present in a single red blood cell and compare it with the cell volume. How tightly are hemoglobin molecules packed in the red blood cell?

(f) In light of your observations in part (e), are the hemoglobin molecules close enough to interact with each other? If so, is it possible that the interaction of hemoglobin S molecules in sickled red cells could affect the shape of the red blood cell?

7.7 Heat dissipation at high metabolic rates. Specific metabolic heat production rate per gram of cell mass is equivalent to 3 nJ per cell per hour [51] or 8×10^{-13} W per cell. In a low cell density culture this energy will dissipate from the cell in a fraction of a millisecond ($\sim$0.01 ms). Estimate the volumetric rate of heat generation in a fermenter with $> 10^8$ cells per milliliter.

B.8   Stoichiometric structure

8.1 Using Eqs (8.2)–(8.5), the rate constants defined below Eq. (8.5) and the nonzero concentrations for the adenosines, redo the simulations and plots in Figure 8.5 by perturbing the steady state with a 50% reduction in the use of ATP; i.e., shift $k_{use} \to 0.5 \times 6.25 = 3.125$. Then interpret the results. Note that Figure 8.5a is a phase portrait of v_{use} versus v_{form}, where the pre-perturbation steady state, $t = 0$, and $t \to \infty$ positions are all indicated.

8.2 Consider steady state in the buffered version of the energy charge (Eq. (8.8) and Figure 8.6) found after simulating using rate constants and nonzero metabolite concentrations defined below Eqs (8.5), (8.12), and (8.13). For each of the following perturbations from steady-state, plot energy charge (Eq. (8.7)) and buffer charge ((8.12), left side) against time as in Figure 8.7, explain the motion of the charges, and identify the key fluxes affected by the perturbation.

(i) ATP use rate reduced by 50%, i.e., $k_{use} = 0.5 \times 6.25 = 3.125$.

(ii) BP $= 0$, B $= 1.0$ mM; all other initial steady-state values the same.

(iii) $K_{buff} = 0.1$.

8.3 Set up the simulation shown in Figure 8.10 using rate constants and metabolite concentrations defined below Eqs (8.5) and (8.13) as initial values to find the steady state before perturbation.

8.4 The synthesis of AMP typically happens through the use of ATP to phosphorylate adenosine (A):

$$A + ATP \to AMP + ADP \tag{B.16}$$

and thus $v_{\text{form}}^{\text{AMP}}$ is not a constant but dependent on ATP. A is typically imported from plasma:

$$v_{\text{form}}^{\text{AMP}} = k_{\text{form}}^{\text{AMP}} \text{ATP} \times \text{A}, \quad \text{and} \quad v_{\text{form}}^{\text{A}} = b_1 \qquad (B.17)$$

(i) Simulate the response of this system perturbed by a 50% increase in the ATP load rate constant, using the rate constants and metabolite concentrations defined below Eqs (8.5) and (8.13) as initial values to find the steady state before perturbation.

(ii) The net effect of adding this is to shift the "cost structure" of the network, as AMP now has been formed using a high-energy bond and thus has a value of one such bond to the system. How would you now define "capacity," "occupancy," and "charge?"

8.5 Set up the simulation that is shown in Figure 8.14.

(i) Plot the occupancy (2ATP + ADP) versus the capacity (ATP + ADP + AMP) in a phase portrait format. Use this plot to interpret what happens to each quantity on the two principal time scales. Interpret the dynamics the stoichiometric structure produces given the rate constants in this case.

(ii) Study the pool $x_1 + 2x_2 + 2\text{ATP} + \text{ADP}$. Obtain the differential equation that describes this pool. What does this pool represent? Study its dynamics in the above simulation.

B.9 Regulation as elementary phenomena

9.1 Look through the chapters in your biochemistry textbook that cover glycolysis, the pentose pathway, and the tricarboxylic acid cycle. How many examples of regulation can you find that are shown in Figure 9.3?

9.2 Why does the inflection point on Hill rate law (recall Eq. (5.27)) not predict a minimal eigenvalue (see Section 9.4.1).

9.3 Redo the analyses of Section 9.4.1 with the symmetry rate law

$$v_1(x) = \frac{V_m}{1 + L(1 + x/K)^2}. \qquad (B.18)$$

(i) Follow the derivation from Eqs (9.9)–(9.11), defining a cubic equation in terms of L and dimensionless variables a and χ, where neither of the latter variables is dependent on L.

(ii) Follow Eq. (9.14) to define a concentration derivative with dimensionless time constant π in terms of a, χ, and L.

(iii) Define eigenvalue λ in terms of χ_{ss} and L following Eq. (9.15).

(iv) For $L = 1$, redo Figure 9.4 for a ranging from 0 to 10 and Figure 9.5 for the values of a indicated.

(v) For $L = 0.05$, redo Figure 9.4 for a ranging from 0 to 10 and Figure 9.5 for the values of a indicated.

9.4 Redo the analysis in Section 9.4 without using the QEA on the Hill rate law for (i) inhibition and (ii) activation. Use $v = 1$ as an example. How much does the QEA affect the results presented in Section 9.4?

9.5 Repeat the transient response shown in Figure 9.6d for $a = 0.2, 0.175, 0.15, 0.125$, and 0.1. Graph the results on one plot of χ versus time. Show the points in the a, α plane. Interpret the results.

9.6 Repeat the computations of a as a function of χ_{ss} with $\alpha = 10$ and $v = 3$ in Figure 9.6b. At the same time, compute the values for λ. Plot both a and λ as a function of χ_{ss}. Interpret the result. Why do the transient computations never settle down in the middle steady state?

9.7 If the removal rate $v_2(x)$ in Eq. (9.1) is of the Michaelis–Menten form,

$$v_2(x) = \frac{V_{m_2} X}{K_m + x},$$
(B.19)

then the steady-state equation can have zero, one, two, or three roots, giving rise to a much richer pattern of multiple steady states.

(i) Scale the equations so that time is scaled to K_m/V_{m_2}, the first-order turnover rate of v_2, and concentration to K_m. Identify and characterize the dimensionless groups.

(ii) Write down the steady-state equation, with $v_1(x) = v_{m_1}/[1 + (x/K)^v]$.

(iii) Find the expression for λ.

(iv) Combine the steady-state equation and the condition $\lambda = 0$.

(v) Find the conditions for zero, one, two, or three roots of the steady-state equation.

9.8 The Goodwin equations. In the early 1960s the first kinetic models were formulated to describe feedback inhibition. They were based on the schema

$$X_1 \xrightarrow{v_1(x_n)} X_2 \xrightarrow{k_2} X_3 \xrightarrow{k_3} X_4 \xrightarrow{k_4} \times \xrightarrow{k_{n-1}} X_n \xrightarrow{k_n},$$
(B.20)

where

$$v_1 = \frac{V_m}{1 + (x_n/K)^v}.$$
(B.21)

It has been shown that this control system can exhibit sustained oscillatory behavior. For this to happen, a critical relationship between v

and n had to be satisfied:

$$v \geq \frac{1}{\cos^n(\pi/n)}. \tag{B.22}$$

(i) Compute the relationship between the chain length n and the degree of cooperativity v.

(ii) Determine the steady-state equations.

(iii) Simulate the dynamic response for $n = 5$ and $k_2 = k_3 = k_4 = k_5 = 1.0$. Vary v from sub- to super-critical based on Eq. (B.22).

9.9 Verify Eq. (9.44). **Hint:** Solve Eqs (9.37)–(9.43) combined with the enzyme conservation constraint.

9.10 Simulate the control loop described in Section 9.5. Use the rate constants given in Figure 9.8, with initial conditions $x_1 = x_6 = 0$ and all other $x_i = 0$, to find the steady state. For each of the following perturbations from steady state, plot the metabolite concentrations' responses together against log-scale time in a single graph, then explain why the perturbation had the observed effects on the control loop.

(i) Increase x_5 by a factor of 10 from its steady-state value.

(ii) Increase x_1 by a factor of 10 from its steady-state value.

9.11 Simulate the regulation of protein synthesis with the dimer and tetramer as the regulatory enzymes by integrating the two extensions of the basic feedback loop described in Section 9.6.

(i) Verify that the tetramer is a more effective mechanism than the dimer for rejecting disturbance imposed on the biochemical system even when feedback of protein synthesis is included.

(ii) Study the reset of the fraction of inactive enzyme described in Eqs (9.52) and (9.53) and how the new steady state is determined.

(iii) Try to find the parameter values that lead to very effective disturbance rejection.

9.12 Consider the prototypic feedback loop of Eqs (9.29)–(9.35).

(i) Write down **S** for this system.

(ii) Compute the rank of **S**.

(iii) Show that the rank deficiency corresponds to the mass balance of Eq. (9.36).

(iv) Find the vector that spans the left null space.

9.13 Build a model of feedback inhibition by combining the regulation of protein synthesis and tetramer symmetry-type model for the inhibition of v_1.

9.14 Consider the basic feedback inhibition schema considered in Section 9.6.1. The disturbance rejection problem can be improved if the end product of the pathway, x_5, accelerates the degradation of x_1 through activation of v_0. Consider an activation-type Hill rate law for this purpose.

  (i) Pick the parameters of this rate law such that it gives the same steady-state flux as the initial state of the simulation in Figure 9.10.
  (ii) How many such choices are there?
  (iii) Judiciously pick parameter sets for the Hill rate law and simulate the effects of this activation on the ability to reject the 10-fold increase in b_1. Find both "poor" and "good" parameter values to improve the rejection problem.

B.10 Glycolysis

10.1 Implement the model of glycolysis discussed in this chapter and reproduce the main dynamic simulation presented in the chapter.

10.2 From Problem 10.1, plot the concentration of pyruvate as a function of time. Discuss and interpret this response.

10.3 From Problem 10.1, change the concentration of pyruvate to a higher value in the physiological range 0.06 to 0.11 mM. Discuss and interpret this response. Examine the mass action ratio for LDH.

10.4 From Problem 10.1, change the concentration of lactate into a lower value in the physiological range 0.4 to 1.8 mM. Discuss and interpret this response. Examine the mass action ratio around LDH.

10.5 Simulate the response to a 50% increase in the rate of NADH utilization. Interpret the response as done in this chapter.

10.6 Evaluate the consequences of the conservation of total phosphate.
  (i) Simulate the response of the glycolytic system to a 5% increase in the glucose uptake rate. Observe what happens to the fraction of phosphate in the recycle pool. Gradually increase the glucose uptake rate. What happens?
  (ii) Add a reaction $v_P = k_P(P_i - P_{i,\text{plasma}})$ where $P_{i,\text{plasma}} = 2.5$ and k_P is 1/s and repeat the simulations from part (i).

10.7 Analogous to the previous problem, evaluate the consequences of the conservation of total NADH (N_{tot}).

B.11 Coupling pathways

11.1 Consider the production of E4P by the pentose pathway. Add a removal reaction to Table 11.3 for E4P and add a corresponding column to the stoichiometric matrix.

(i) Evaluate the rank of **S** and the dimensions of the null spaces. How do they change?

(ii) Use first-order removal kinetics:

$$v_{E4Pr} = k_{E4Pr}E4P. \tag{B.23}$$

Put the rate constant for the removal rate to zero, $k_{E4Pr} = 0$. The steady state should be unaltered. Then change k_{E4Pr} to a finite number. Use $E4P_{stst} = 0.0127$ mM to calibrate the removal flux. First remove 1% of the incoming pentose pathway flux as E4P.

(iii) Then increase this percentage to a more significant number. Can you exceed the input flux into the oxidative branch of the pentose pathway through v_{G6PDH}?

11.2 Consider a 10% drop in the rate of GSH utilization.

(i) Simulate the dynamic response using the Mathematica workbook provided. Plot the concentration of GL6P and GO6P as a function of time. What do you observe?

(ii) Analyze the result from by re-examining how the fluxes on GL6P and GO6P can be balanced in the light of a constant input into the pathway. Add a dependency of the input to NADP as

$$v_{G6Pin} = k_{G6Pin}NADH. \tag{B.24}$$

Pick a value for k_{G6Pin} such that the input is 0.21 mM/h initially. Now repeat the simulation.

B.12 Forming integrated networks

12.1 Simulate the response of the AMP sub-network to a 10% sudden drop at $t = 0$ in the AMP concentration.

(i) Show that the response is the inverse of the response shown in Figure 12.3.

(ii) Discuss the relative fluxes of the degradation pathways and the biosynthetic pathways.

12.2 Simulate the response of the integrated model to an increase in the rate of GSH utilization as done in Figure 11.7 for the coupled glycolysis and pentose pathway model. Compare the difference in the two responses and discuss the effects (or the lack thereof) of the addition of the AMP sub-network. Do pay attention to the fact that R5P is a common metabolite in the pentose pathway and the AMP sub-network.

12.3 The adenine phopshoribosyl transferase reaction does actually produce pyrophosphate, PP_i and not $2P_i$ as indicated in reaction 41 in

Table 12.3. It is then hydrolyzed by the reaction

$$PP_i + H_2O \rightarrow 2P_i. \tag{B.25}$$

(i) Add this reaction to the stoichiometric matrix. Make sure that you have balanced the hydrogen properly.
(ii) Simulate the response to the 50% increase in the rate of use of ATP. Show that this addition does not change the dynamic response.

12.4 Make a list of the additions to the integrated model discussed in Section 12.3.
(i) Evaluate and describe what is needed to add each of these features into the integrated model.
(ii) Review the literature on red blood cell metabolism and determine if the list of additional features described in Section 12.3 is complete and if there are additional metabolic effects that should be added into a whole-cell model.

B.13 Hemoglobin

13.1 Consider to add the half-life of hemoglobin and the red cell as a multi-month-long physiological time scale. Discuss why and why not to add this into the model.

13.2 Based on the simulators in this chapter, compute the total oxygen delivered to tissues in 1 h. Look up the oxygen consumption rate in a 70 kg human in the literature for comparison.

13.3 For the hemoglobin sub-network alone, combine the normal oscillatory oxygen concentration and the shift in the average partial pressure of oxygen. Compute and plot the oxygen delivered as a function of time, as well as the plot that corresponds to Figure 13.3c.

13.4 Simulate the dynamic response of the glycolytic/hemoglobin model of this chapter to an increased ATP rate of use as in Chapter 10 and compare the results with those presented in this chapter.

13.5 Couple the hemoglobin model to the glycolytic/pentose pathway model of Chapter 11. Repeat the two simulations covered in this chapter as well as the ATP load of Chapter 11. Discuss the results and compare them with the simpler constituent models.

13.6 Further to Problem 13.4, look up the oxidation rate of iron in hemoglobin ($Fe^{2+} \rightarrow Fe^{3+}$) and add this oxidation load to the model. **Hint:** About 1.5% of the iron in hemoglobin is oxidized daily.

B.14 Regulated enzymes

14.1 Examine the effect of changing the values of the three kinetic parameters that characterize PFK shown in Figure 14.3.

  (i) Find numerical values for the three rate constants such that r_R is lower than 0.89. Try to make r_R closer to 0.50.

  (ii) Redo simulations associated with 50% increase in the rate of ATP utilization. Compare the two dynamic responses that result from this change.

References

[1] http://www.rcsb.org/pdb/explore/explore.do?structureId=1B8P.

[2] http://xpdb.nist.gov/enzyme_thermodynamics/.

[3] http://users.rcn.com/jkimball.ma.ultranet/BiologyPages/G/GenomeSizes. html.

[4] http://www.altitude.org/calculators/oxygencalculator/oxygencalculator. htm.

[5] http://www.magnet.fsu.edu/education/tutorials/java/ignitioncoil/index. html.

[6] K.R. Albe, M.H. Butler, and B.E. Wright. Cellular concentrations of enzymes and their substrates. *Journal of Theoretical Biology*, 143:163–195, 1990.

[7] B. Alberts, D. Bray, J. Lewis, M. Raff, K. Roberts, and J.D. Watson. *Molecular Biology of the Cell*. Garland Publishing, Inc., New York, 1983.

[8] R.A. Alberty. *Thermodynamics of Biochemical Reactions*. Wiley-Interscience, Hoboken, NJ, 2003.

[9] R.A. Alberty. *Biochemical Thermodynamics: Applications of Mathematica*. Wiley-Interscience, Hoboken, NJ, 2006.

[10] W.J. Albery and J.R. Knowles. Evolution of enzyme function and development of catalytic efficiency. *Biochemistry*, 15:5631–5640, 1976.

[11] W.J. Albery and J.R. Knowles. Efficiency and evolution of enzyme catalysis. *Angewandte Chemie, International Edition in English*, 16:285–293, 1977.

[12] O. Alter, P.O. Brown, and D. Botstein. Singular value decomposition for genome-wide expression data processing and modeling. *Proceedings of the National Academy of Sciences of the United States of America*, 97:10101–10106, 2000.

[13] D.E. Atkinson. The energy charge of the adenylate pool as a regulatory parameter. interaction with feedback modifiers. *Biochemistry*, 7:4030, 1968.

[14] D.E. Atkinson. *Cellular Energy Metabolism and its Regulation*. Academic Press, New York, 1977.

[15] J.E. Bailey. Mathematical modeling and analysis in biochemical engineering: past accomplishments and future opportunities. *Biotechnology Progress*, 14:8–20, 1998.

[16] I. Baroli, A.D. Do, T. Yamane, and K.K. Niyogi. Zeaxanthin accumulation in the absence of a functional xanthophyll cycle protects *Chlamydomonas reinhardtii* from photooxidative stress. *The Plant Cell*, 15:992–1008, 2003.

[17] E. Beutler and W.J. Williams. *Williams Hematology*. McGraw Hill Health Professions Division, 2001.

[18] F.R. Blattner, G. Plunkett III, C.A. Bloch *et al.* The complete genome sequence of *Escherichia coli* K-12. *Science*, 277(5331):1453–74, 1997.

[19] J.R. Bowen, A. Acrivos, and A.K. Oppenheim. Singular perturbation refinement to quasi-steady state approximation in chemical kinetics. *Chemical Engineering Science*, 18:177–188, 1963.

[20] P.D. Boyer, H. Lardy, and K. Myrback, editors. *The Enzymes*, pp. 1–48. Academic Press, New York, 1959.

[21] H. Bremer and P.P. Dennis. Modulation of chemical composition and other parameters of the cell by growth rate. In *Escherichia coli and Salmonella typhimurium: Cellular and Molecular Biology*. ASM Press, Washington, DC, 1996.

[22] G.E. Briggs and J.B.S. Haldane. A note on the kinetics of enzyme action. *Biochemical Journal*, 19:338–339, 1925.

[23] R. Burns, T. Friedmann, W. Driever, M. Burrascano, and J.K. Yee. Vesicular stomatitis virus G glycoprotein pseudotyped retroviral vectors: concentration to very high titer and efficient gene transfer into mammalian and non-mammalian cells. *Proceedings of the National Academy of Sciences of the United States of America*, 90:8033–8037, 1993.

[24] A.K. Christensen, L.E. Kahn, and C.M. Bourne. Circular polysomes predominate on the rough endoplasmic reticulum of somatotropes and mammotropes in the rat anterior pituitary. *American Journal of Anatomy*, 178:1–10, 1987.

[25] A.S. Chuck, M.F. Clarke, and B.O. Palsson. Retroviral infection is limited by Brownian motion. *Human Gene Therapy*, 7:1527–1534, 1996.

[26] W.W. Cleland. What limits the rate of an enzyme catalyzed reaction? *Accounts of Chemical Research*, 8:145–151, 1981.

[27] P.P. Dennis and H. Bremer. Differential rate of ribosomal protein synthesis in *Escherichia coli* B/r. *Journal of Molecular Biology*, 84:407–422, 1974.

[28] W. Dudzinska, A.J. Hlynczak, E. Skotnicka, and M. Suska. The purine metabolism of human eythrocytes. *Biochemistry*, 71:467–475, 2006.

[29] J.S. Edwards and B.O. Palsson. Systems properties of the *Haemophilus influenzae* Rd metabolic genotype. *Journal of Biological Chemistry*, 274:17410–17416, 1999.

[30] J.S. Edwards and B.O. Palsson. The *Escherichia coli* MG1655 *in silico* metabolic genotype; its definition, characteristics, and capabilities.

Proceedings of the National Academy of Sciences of the United States of America, 97:5528–5523, 2000.

[31] M.A. Eiteman, S.A. Lee, R. Altman, and E. Altman. A substrate-selective co-fermentation strategy with *Escherichia coli* produces lactate by simultaneously consuming xylose and glucose. *Biotechnology and Bioengineering*, 102:822–827, 2009.

[32] I. Famili, R. Mahadevan, and B.O. Palsson. *k*-Cone analysis: determining all candidate values for kinetic parameters on a network-scale. *Biophysical Journal*, 88:1616–1625, 2005.

[33] I. Famili and B.O. Palsson. Systemic metabolic reactions are obtained by singular value decomposition of genome-scale stoichiometric matrices. *Journal of Theoretical Biology*, 224:87–96, 2003.

[34] A.M. Feist, C.S. Henry, J.L. Reed *et al.* A genome-scale metabolic reconstruction for *Escherichia coli* K-12 MG1655 that accounts for 1260 ORFs and thermodynamic information. *Molecular Systems Biology*, 3:121, 2007.

[35] A.M. Feist, M.J. Herrgard, I. Thiele, J.L. Reed, and B.O. Palsson. Reconstruction of biochemical networks in microbial organisms. *Nature Reviews Microbiology*, 7(2):129–143, 2009.

[36] A.M. Feist and B.O. Palsson. The growing scope of applications of genome-scale metabolic reconstructions using *Escherichia coli*. *Nature Biotechnology*, 26:659–667, 2008.

[37] H.M Gilmore and R. Wall. Nuclear RNA precursors in the processing pathway to MOPC 21 kappa light chain messenger RNA. *Journal of Molecular Biology*, 135:879–891, 1979.

[38] D.S. Goodsell. *The Machinery of Life*. Springer-Verlag, New York, 1993.

[39] H. Gutfreund. *Enzymes: Physical Principles*. John Wiley & Sons, London, 1972.

[40] E.H. Harris. *The Chlamydomonas Sourcebook: A Comprehensive Guide to Biology and Laboratory Use*. Academic Press, 1989.

[41] F.H. Heineken, H.M. Tsuchyia, and R. Aris. On the mathematical status of pseudo steady-state hypothesis in biochemical kinetics. *Mathematical Biosciences*, 1:95–113, 1967.

[42] R. Heinrich, S.M. Rapoport, and T.A. Rapoport. Metabolic regulations and mathematical models. *Progress in Biophysics and Molecular Biology*, 32:1–82, 1977.

[43] V. Henri. *Lois Generales de l'action des Diastases*. Hermann, 1903.

[44] C.S. Henry, M.D. Jankowski, L.J. Broadbelt, and V. Hatzimanikatis. Genome-scale thermodynamic analysis of *Escherichia coli* metabolism. *Biophysical Journal*, 90(4):1453–61, 2006.

[45] C.S. Henry, L.J. Broadbelt, and V. Hatzimanikatis. Thermodynamics-based metabolic flux analysis. *Biophysical Journal*, 92:1792–1805, 2007.

[46] A.V. Hill. The possible effects of the aggregation of the molecules of haemoglobin on its dissociation curves. *Journal of Physiology*, 40:4–7, 1910.

[47] A.V. Hill. Specific ligand-induced association of an enzyme. a new model of dissociating allosteric enzyme. *Journal of Physiology*, 21:227–245, 1983.

[48] F.A. Hommes. The integrated Michaelis–Menten equation. *Archives of Biochemistry and Biophysics*, 96:28–31, 1962.

[49] D.R. Hyduke and B.O. Palsson. Towards genome-scale signalling-network reconstructions. *Nature Reviews Genetics*, 11:297–307, 2010.

[50] J.L. Ingraham, O. Maaloe, and F.C. Neidhardt. *Growth of the Bacterial Cell*. Sinauer Associates, Inc., Sutherland, 1983.

[51] Y. Ishikawa and M. Shoda. Calorimetric analysis of *Escherichia coli* in continuous culture. *Biotechnology and Bioengineering*, 25:1817–1827, 1983.

[52] N. Jamshidi and B.O. Palsson. Formulating genome-scale kinetic models in the post-genome era. *Molecular Systems Biology*, 4:171, 2008.

[53] N. Jamshidi and B.O. Palsson. Top-down analysis of temporal hierarchy in biochemical reaction networks. *PLoS Computational Biology*, 4:e1000177, 2008.

[54] N. Jamshidi and B.O. Palsson. Mass action stoichiometric simulation models: incorporating kinetics and regulation into stoichiometric models. *Biophysical Journal*, 98:175–185, 2010. DOI: 10.1016/j.bpj.2009.09.064.

[55] N. Jamshidi, S.J. Wiback, and B.O. Palsson. *In silico* model-driven assessment of the effects of single nucleotide polymorphisms (SNPs) on human red blood cell metabolism. *Genome Research*, 12:1687–1692, 2002.

[56] K. Jantama, X. Zhang, J.C. Moore, K.T. Shanmugam, S.A. Svoronos, and L.O. Ingram. Eliminating side products and increasing succinate yields in engineered strains of *Escherichia coli* C. *Biotechnology and Bioengineering*, 101:881–893, 2008.

[57] M. Javanmardian and B.O. Palsson. High-density photoautotrophic algal cultures: design, construction, and operation of a novel photobioreactor system. *Biotechnology and Bioengineering*, 38:1182–1189, 1991.

[58] A. Joshi. Integrated metabolic dynamics in the human red blood cell. PhD thesis, University of Michigan, 1988.

[59] A. Joshi and B.O. Palsson. Metabolic dynamics in the human red cell. Part I – a comprehensive kinetic model. *Journal of Theoretical Biology*, 141:515–528, 1989.

[60] A. Joshi and B.O. Palsson. Metabolic dynamics in the human red cell. Part III – metabolic reaction rates. *Journal of Theoretical Biology*, 142:41–68, 1990.

[61] A. Joshi and B.O. Palsson. Metabolic dynamics in the human red cell. Part II – interactions with the environment. *Journal of Theoretical Biology*, 141:529–545, 1989.

[62] A. Joshi and B.O. Palsson. Metabolic dynamics in the human red cell. Part IV – data prediction and some model computations. *Journal of Theoretical Biology*, 142:69–85, 1990.

[63] K. Schwerzmann, L.M. Cruz-Orive, R. Eggman, A. Sänger, and E.R. Weibel. Molecular architecture of the inner membrane of mitochondria from rat

liver: a combined biochemical and stereological study. *Journal of Cell Biology*, 102:97–103, 1986.

[64] K.J. Kauffman, J.D. Pajerowski, N. Jamshidi, B.O. Palsson, and J.S. Edwards. Description and analysis of metabolic connectivity and dynamics in the human red blood cell. *Biophysical Journal*, 83:646–662, 2002.

[65] E. Klipp, B. Nordlander, R. Kruger, P. Gennemark, and S. Hohmann. Integrative model of the response of yeast to osmotic shock. *Nature Biotechnology*, 23:975–982, 2005.

[66] B. Kok. Efficiency of photosynthesis. In W. Ruhland, editor, *Encyclopedia of Plant Physiology*, pp. 566–633. Springer, 1960.

[67] W.N. Konings and H. Veldkamp. Energy transduction and solute transport mechanisms in relation to environments occupied by microorganisms. In J.H. Slater, R. Whittenbury, and J.W.T. Wimpenny, editors, *Microbes in their Natural Environments*, pp. 153–186. Cambridge University Press, 1983.

[68] J. Koolman and K.H. Roehm. *Color Atlas of Biochemistry*. Thieme, New York, 2005.

[69] C.G. Lee and B.O. Palsson. High-density algal photobioreactors using light emitting diodes. *Biotechnology and Bioengineering*, 44:1161–1167, 1994.

[70] S.E. Leney and J.M. Tavare. The molecular basis of insulin-stimulated glucose uptake: signalling, trafficking and potential drug targets. *Journal of Edocrinology*, 203:1–18, 2002.

[71] E.M. Levine. Protein turnover in *Escherichia coli* as measured with an equilibration apparatus. *Journal of Bacteriology*, 90:1578–1588, 1965.

[72] C.C Lin and L.A Segel. *Mathematics Applied to Deterministic Problems in the Natural Sciences*. Macmillan, New York, 1974.

[73] W.R. Loewenstein. *The Touchstone of Life*. Oxford University Press, Oxford, 1999.

[74] C.J. Masters. Metabolic control and the microenvironment. *Current Topics in Cellular Regulation*, 12:75–105, 1977.

[75] M.L. Mavrovouniotis. Estimation of standard Gibbs energy changes of biotransformations. *Journal of Biological Chemistry*, 266(22):14440–14445, 1991.

[76] M. McMullin. The molecular basis of disorders of red cell enzyme. *Journal of Clinical Pathology*, 52:241–244, 1999.

[77] M. Mehta, M. Haripalsingh, and S.S. Sonawat. Malaria parasite-infected erythrocytes inhibit glucose utilization in uninfected red cells. *FEBS Letters*, 579(27):6151–6158, 2005.

[78] W. Meiske. An approximate solution of the Michaelis–Menten mechanism for quasi-steady state and quasi-equilibrium. *Mathematical Biosciences*, 42:63–71, 1978.

[79] A. Melis, L. Zhang, M. Forestier, M.L. Ghirardi, and M. Seibert. Sustained photobiological hydrogen gas production upon reversible inactivation of oxygen evolution in the green alga *Chlamydomonas reinhardtii*. *Plant Physiology*, 122:127–135, 2000.

[80] L. Michaelis and M. Menten. Die kinetik der invertinwirkung. *Biochemische Zeitschrift*, 49:333–369, 1913.

[81] P. Mitchell and J. Moyle. Translocation of some anions cations and acids in rat liver mitochondria. *European Journal of Biochemistry*, 9:149–155, 1969.

[82] J. Monod, J. Wyman, and J.P. Changeaux. On the nature of allosteric transitions: a plausible model. *Journal of Molecular Biology*, 12:88–118, 1965.

[83] J. Myers. On the algae: thoughts about physiology and measurements of efficiency. In P.G. Falkowski, editor, *Primary Productivity in the Sea*, pp. 1–16. Plenum Press, 1980.

[84] F.C. Neidhardt, editor. *Escherichia coli and Salmonella: Cellular and Molecular Biology*. ASM Press, Washington, DC, 1996.

[85] D.G. Nicholls. Brown adipose tissue mitochondria. *Biochimica Biophysica Acta*, 549:1–29, 1979.

[86] M.A. Oberhardt, B.O. Palsson, and J.A. Papin. Applications of genome-scale metabolic reconstructions. *Molecular Systems Biology*, 5:320, 2009.

[87] D.A. Okar, A.J. Lange, À. Manzano, A. Navarro-Sabatè, L. Riera, and R. Bartrons. PFK-2/FBPase-2: maker and breaker of the essential biofactor fructose-2,6-bisphosphate. *Trends in Biochemical Sciences*, 26(1):30–35, 2001.

[88] B.O. Palsson, J.C. Liao, and E.N. Lightfoot. Interpretation of biochemical dynamics using modal analysis. Presented at the *Annual AIChE Meeting*, San Francisco, CA, 1984.

[89] B.O. Palsson. *Systems Biology: Properties of Reconstructed Networks*. Cambridge University Press, New York, 2006.

[90] B.O. Palsson. *Systems Biology: Volume III*. Cambridge University Press, Cambridge. In preparation.

[91] B.O. Palsson and T. Groshans. Mathematical modelling of dynamics and control in metabolic networks. Part VI. Dynamic bifurcations in single biochemical control loops. *Journal of Theoretical Biology*, 131:43–53, 1988.

[92] B.O. Palsson, R. Jamier, and E.N. Lightfoot. Mathematical modelling of dynamics and control in metabolic networks. Part II. Simple dimeric enzymes. *Journal of Theoretical Biology*, 111:303–321, 1984.

[93] B.O. Palsson and E.N. Lightfoot. Mathematical modelling of dynamics and control in metabolic networks. Part I. On Michaelis–Menten kinetics. *Journal of Theoretical Biology*, 111:273–302, 1984.

[94] J.A. Papin, N.D. Price, S.J. Wiback, D.A. Fell, and B.O. Palsson. Metabolic pathways in the post-genome era. *Trends in Biochemical Sciences*, 28:250–258, 2003.

[95] K.F. Petersen and G.I. Shulman. Pathogenesis of skeletal muscle insulin resistance in type 2 diabetes mellitus. *American Journal of Cardiology*, 90:11G–18G, 2002.

[96] S.J. Pirt, Y.K. Lee, A. Richmond, and M.W. Pirt. The photosynthetic efficiency of *Chlorella* biomass growth with reference to solar energy utilization. *Journal of Chemical Technology and Biotechnology*, 30:25–34, 1980.

[97] V.A. Portnoy, M.J. Herrgard, and B.O. Palsson. Aerobic fermentation of D-glucose by an evolved cytochrome oxidase-deficient *Escherichia coli* strain. *Applied Environmental Biology*, 74:7561–7569, 2008.

[98] M. Potter. Immunoglobulin-producing tumors and myeloma proteins of mice. *Physiological Review*, 52:631–719, 1972.

[99] J.M. Pratt, J. Pretty, I. Riba-Garcia *et al.* Dynamics of protein turnover, a missing dimension in proteomics. *Molecular and Cellular Proteomics*, 1:579–591, 2002.

[100] J.L. Reed, I. Famili, I. Thiele, and B.O. Palsson. Towards multidimensional genome annotation. *Nature Reviews Genetics*, 7(2):130–41, 2006.

[101] J.G. Reich and E.E. Sel'kov. *Energy Metabolism of the Cell*. Academic Press, New York, 1981.

[102] M. Riley, T. Abe, M.B. Arnaud, *et al. Escherichia coli* K-12: a cooperatively developed annotation snapshot-2005. *Nucleic Acids Research*, 34(1):1–9, 2006.

[103] C. Rosenow, R. Mukherjee Saxena, M. Durst, and T.R. Gingeras. Prokaryotic RNA preparation methods useful for high density array analysis: comparison of two approaches. *Nucleic Acids Research*, 29:E112, 2001.

[104] S.A. Ross, M.X. Zhang, and B.R. Selman. Role of the *Chlamydomonas reinhardtii* coupling factor 1 γ-subunit cysteine bridge in the regulation of ATP synthase. *Journal of Biological Chemistry*, 270:9813–9818, 1995.

[105] H. Rottenberg and T. Grunwald. Determination of pH in chloroplasts. 3. Ammonium uptake as a measure of pH in chloroplasts and sub-chloroplast particles. *European Journal of Biochemistry*, 25:71–74, 1972.

[106] M.A. Savageau. Optimal design of feedback control by inhibition. *Journal of Molecular Evolution*, 29:139–156, 1974.

[107] J.M. Savinell and B.O. Palsson. Network analysis of intermediary metabolism using linear optimization: II. Interpretation of hybridoma cell metabolism. *Journal of Theoretical Biology*, 154:455–473, 1992.

[108] J.M. Savinell, G.M. Lee, and B.O. Palsson. On the orders of magnitude of epigenic dynamics and monoclonal antibody production. *Bioprocess Engineering*, 4:231–234, 1989.

[109] U. Schibler, K.B. Marcu, and R.P. Perry. The synthesis and processing of the messenger RNAs specifying heavy and light chain immunoglobulins in MPC-11 cells. *Cell*, 15:1495–1509, 1978.

[110] M.C. Schrader, C.J. Eskey, V. Simplaceanu, and C. Ho. A carbon-13 nuclear magnetic resonance investigation of the metabolic fluxes associated with glucose metabolism in human erythrocytes. *Biochimica Biophysica Acta.*, 1182(2):162–178, 1993.

[111] S. Schuldiner, E. Padan, H. Rottenbergy, Z. Gromet-Elhanan, and M. Avron. Delta pH and membrane potential in bacterial chromatophores. *FEBS Letters*, 49:174–177, 1974.

[112] R.C. Seagrave. *Biomedical Applications of Heat and Mass Transfer*. Iowa State Press, Ames, 1971.

[113] I.H. Segal. *Enzyme Kinetics: Behavior and Analysis of Rapid Equilibrium and Steady-State Enzyme Systems*. Wiley, New York, 1975.

[114] M.C. Sorgato, S.J. Ferguson, and D.B. Kell. On the current–voltage relationships of energy-transducing membranes: submitochondrial particles. *Biochemical Society Transcripts*, 6:1301–1302, 1978.

[115] P.A. Srere. Enzyme concentrations in tissue. *Science*, 158:936–937, 1967.

[116] P.A. Srere. Enzyme concentrations in tissue II. *Biochemical Medicine*, 4:43–46, 1970.

[117] I. Thiele and B.O. Palsson. A protocol for generating a high-quality genome-scale metabolic reconstruction. *Nature Protocols*, 5:93–121, 2010.

[118] J.J. Tyson. Periodic enzyme synthesis and oscillatory repression: why is the period of oscillation so close to the cell cycle time? *Journal of Theoretical Biology*, 103:313–328, 1983.

[119] N. vanThoai and J. Roche. Phosphagens of marine animals. *Annals of the New York Academy of Sciences*, 90:923–928, 1960.

[120] A. Varma and B.O. Palsson. Metabolic capabilities of *Escherichia coli*. I. Synthesis of biosynthetic precursors and cofactors. *Journal of Theoretical Biology*, 165:477–502, 1993.

[121] J.C. Venter. *E. coli* sequencing. *Science*, 267:172–174, 1995.

[122] R. Wagner. *Transcription Regulation in Prokaryotes*. Cambridge University Press, Cambridge, 2000.

[123] R.T. Watson and J.E. Pessin. Bridging the GAP between insulin signaling and GLUT4 translocation. *Trends in Biochemical Sciences*, 31:215–222, 2006.

[124] P.B. Weisz. Diffusion and chemical transformation. *Science*, 179:433–440, 1973.

[125] U. Wittig, M. Golebiewski, R. Kania *et al.* SABIO-RK: integration and curation of reaction kinetics. In U. Leser, F. Naumann, and B. Eckman, editors, *Data Integration in the Life Sciences. Third International Workshop, DILS 2006, Hinxton, UK, July 20–22, 2006. Proceedings*, pp. 94–103. Lecture Notes in Computer Science (Lecture Notes in Bioinformatics), vol. 4075. Springer-Verlag, Berlin, 2006.

[126] L.P. Yomano, S.W. York, S. Zhou, K.T. Shanmugam, and L.O. Ingram *et al.* Re-engineering *Escherichia coli* for ethanol production. *Biotechnology Letters*, 30:2097–2103, 2008.

[127] R. Young and H. Bremer. Polypeptide-chain-elongation rate in *Escherichia coli* B/r as a function of growth rate. *Biochemical Journal*, 160:185–194, 1976.

[128] J. Yuan, W.U. Fowler, E. Kimball, W. Lu, and J.D. Rabinowitz. Kinetic flux profiling of nitrogen assimilation in *Escherichia coli*. *Nature Chemical Biology*, 2(10):529–530, Oct 2006.

[129] Y. Zhu, M.A. Eiteman, R. Altman, and E. Altman. High glycolytic flux improves pyruvate production by a metabolically engineered *Escherichia coli* strain. *Applied Environmental Biology*, 74:6649–6655, 2008.

[130] D. Zilberstein, S. Schuldiner, and E. Padan. Proton electrochemical gradient in *Escherichia coli* cells and its relation to active transport of lactose. *Biochemistry*, 18:669–673, 1979.

Index

2,3-biphosphoglycerate (DPG23), 248

absolute rates, 59
activation energy, 77
active range, 152
aggregate variable, 6, 18, 25, 27, 46, 49, 54, 277
allosteric constant, 91
allosteric site, 90
amino acid synthesis, 205
AMP input, 224
AMP metabolism, 224, 226, 231
AMP output, 224
anabolic redox charge, 28
analyzing dynamic models, 13
Arrhenius' law, 9
assumptions concerning kinetic models, 8
Atkinson's energy charge, 139
attracting fixed point, 46

Bailey's five reasons, 5
bilinear reaction, 58, 134
  connected reversible, 70
  reversible, 62
bilinear term, 65
bias, 152
bifurcation theory, 7
biochemical reaction, 59
biosynthetic precursor, 161
bottom-up approach to models, 7
BRENDA, 120
Brownian motion, 8
Buckingham Pi theorem, 95
buffer molecules, 140
buffering, 140

C. reinhardtii, 123
C. vulgaris, 128
capacity, 27
carrier, 134, 135
catabolic redox charge, 27
cell
  chemical composition, 113
  growth, 128
  protein concentration, 124
cell architecture
  fine, 9
chemical assays, 3
  high-throughput, 3
chemical transformations, 1
  absolute rates, 59
  relative rates
    thermodynamics, 59
  stoichiometry, 59
classical rate law, 24
closed system, 58
coarse-grained description, 18
coenzyme, 133, 134, 135, 151
cofactor, 133, 151
collision probability of molecules, 22
conjugate pools, 199
connectivity, 179
conservation quantity, 60, 62, 63, 69, 71, 72, 75, 81
constant-volume assumption, 9, 34, 35
control signals
  metabolic, 152
    activation, 152
    destabilizing, 156
    eigenvalue, 156
    inhibition, 152
    mass action, 155

stabilizing, 156
time constant, 156
cooperativity, 247
correlation coefficient, 51
correlation matrix, **R**, 50
creative functions, 7

decoupling
dynamic, 67
stoichiometric, 67, 69
degradation pathways, 224
degree of cooperativity, 87
deviation variable, 63, 72
deviation variables, 64
dimensionality of data, 283
dimensionless groups, 84
disassociation equilibrium constant,
82
disequilibrium quantity, 62, 63, 64, 69,
72, 75
disturbance rejection, 153
DNA polymerase, 127
dynamic decoupling, 67
dynamic mass balance, 10, 51
equations of, 10, 41
dynamic models, 13
analyzing, 13
simulating, 13
dynamic states, 6
visualising, 20
dynamic structure of network, 6

E. coli, 111–131, 134–135
E. coli cell, 112, 124, 126
eigenrow, 29
eigenvalue, 29, 155, 157, 158, 159, 277
eigenvector, 29
electron transfer system, 28
electroneutrality, 10, 36, 240
elementary reactions, 21
bilinear, 21
linear, 21
energy charge, 27
enzymatic activity, 76
enzymatic rate laws, 79
enzyme
catalysis, 76
classification, 132
kinetic data, 120
regulated, 85
Enzyme Commission (EC) numbers, 133
enzyme kinetics
Hill mechanism, 85

activation, 293
dissociation constant, 87
time of inhibition, 89
symmetry model, 90
enzyme regulation, 150
Epstein–Barr virus (EBV), 125
equilibrium constant, 23
equilibrium state, 63
computation of, 71
extreme pathway
type 1, 100
type 3, 100

feedback
activation, 155, 158
inhibition, 155, 157, 158, 161
feedforward
activation, 155
inhibition, 155
Fermi problem, 114
flux regulation, 259
fructose-6-phosphate (F6P), 18

gain of regulation, 152
gene expression, 150
gene transcription, 150
genetically determined matrix, 12
genome characteristics, 124
genome size, 125
genome-scale reconstructions, 2
genomically derived matrix, 11
Gibbs free energy, 78, 120
glucose-6-phosphate (G6P), 18
glycolysis, 2, 7, 18, 259
gradient matrix, 11, 31
group contribution method, 120

hematocrit, 245
hemoglobin, 245, 246, 248
human, 246
hemoglobin–oxygen binding, 247
Hessian matrix, 65
hexokinase (HK), 7, 18
hexosephosphate pool (HP), 19
high-energy bonds, 145
Hill kinetics, 85
Hill model, 76
homeostatic state, 7, 97, 99

interconversion of enzymes, 150
intracellular environment, 111

Jacobian matrix, 11, 13, 14, 29

kinetic models
  assumptions behind, 8

ligand binding, 151
linear optimization, 218
linear reaction
  connected reversible, 66
  reversible, 60
linearization, 63, 65, 72
link, *see* chemical transformations
local regulation, 155

mass action kinetics, 21
mass action ratio, 24
mass action stoichiometric simulation
    (MASS), 173
mass conservation, 60
MASS model building procedure,
    275
matrix algebra, 11
mechanistic kinetic models, 7
metabolic flux, 117
metabolic networks, 2
metabolic physiology, 27
metabolic process, 116
metabolic turnover, 121
metabolite concentration, 116
Michaelis constant, 82
Michaelis–Menten constant, 24
Michaelis–Menten equation, 82
Michaelis–Menten kinetics, 80
Michaelis–Menten mechanism, 23, 76, 80,
    83, 85, 106, 292
  open environment, 104
moiety exchange, 134
molecular collision, 22
multiple steady states, 33, 159

net rate constant, 153
net rate of reaction, 23
network
  definitions of, 3
  dynamic structure, 6
  metabolic, 2
  reconstructed, 1
  reconstruction of, 4, 275
  regulatory, 3
  signaling, 2
  topology, 152
network dynamics, 6, 11, 13, 14, 41
node, 1
node map, 10, 184
nonoxidative branch, 204

null space, 31, 51
numerical solver, 43

occupancy, 27
omic data, 276, 277
open system, 97
order of magnitude estimation, 114
osmotic balance, 35
osmotic pressure, 10, 240
oxidative branch, 204

p50, 247, 248
partial pressure of oxygen, 247
participation number, 179
pentose pathway, 204, 205
peptide bond, 127
PERCs, 24, 171, 181, 254, 258, 277
pH dependence, 10
phase portrait, 20, 45, 46, 47, 54, 55, 56, 61,
    83, 88, 94
  dynamic, 48
  tiled, 48, 54
  trajectory, 46
phenotypic function, 128
phosphofructokinase (PFK), 18, 259,
    260
phosphogluco-isomerase (PGI), 18
piano tuners in Chicago, 114
pool transformation matrix, 62
  P, 49
pooled variable, *see* aggregate variable, 6
pooling, 18, 218
pools, 25, 31
  conjugate, 199
  hierarchical formation of, 31
  multi-reaction, 67, 69
power density estimation, 123
power-law kinetics, 22
protonmotive force, 122
pseudo-elementary rate constant, 181, 258,
    277
pseudo-first order rate constant, 24

quasi-equilibrium
  Hill kinetics, 87
quasi-equilibrium assumption, 82, 277
  applicability, 88
quasi-steady-state assumption, 24, 82, 84,
    277

random molecular motion, 8
Rapoport–Luebering shunt, 247, 248, 249,
    251, 254, 257, 258

rate law
    enzymatic, 79
    Hill-type, 156
reaction mechanism, 22, 78
reaction rates, 20
reaction specificity, 76
reaction, distance from equilibrium, 25
reconstructed networks, 1
redox potential, 205, 216
reference scales, 94
regulated enzymes, 85
regulation
    enzymes, 150
    Hill kinetics, 156
    protein synthesis, 165, 166
regulatory networks, 3
relative rates, 59
retrovirus
    half-life, 115
reversible reaction
    linear, 60
    open environment, 100
RNA polymerase, 127

SABIO-RK, 120
salvage pathways, 224
scaling, 94, 95
sensing mechanisms, 2
sensitivity, 152
servo problem, 153
shadow prices, 218
signaling networks, 2
simplifying assumptions, 8
simulating dynamic models, 13
smooth functions, 65
sodium–potassium pump, 240
soft glass intracellular environment, 112
sparse matrix, 11

steady state flux, 100
stickiness number, 84
stoichiometric autocatalysis, 32, 33
stoichiometric coefficients, 11
stoichiometric decoupling, 67, 69
stoichiometric matrix, 1, 11, 31
substrate specificity, 77
substrate uptake rate, 128
symmetry model, 76, 90, 94
system boundary, 97
    inputs and outputs, 98
    internal network, 99
    physical, 97
    real, 97
systems versus component variables, 186

temperature
    constant, 9
temporal decomposition, 7
thermodynamic driver, 62
three-dimensional linear system, 29
time constant, 6, 17, 28, 29, 99
time invariants, 31
time profiles, 21, 44
time-scale decomposition, 26, 27, 28, 47
time scales, 28
    multiple, 31
top-down approach to models, 8
tracers, 99
transcription factors, 150
transient response, 19
transition state, 77
transitions, 7, 19

vacancy variable, 27
visualizing dynamic states, 20

window of observation, 28